BASIC MICROBIOLOGY TECHNIQUES

BASIC MICROBIOLOGY TECHNIQUES

Fourth Edition
Revised Printing

Susan G. Kelley, M.D., Ph.D.

Frederick J. Post, Ph.D.

PUBLISHING COMPANY, INC.
P.O. BOX 68
BELMONT, CALIFORNIA 94002
TEL: (650) 591-3505
FAX: (650) 591-3898

Star Publishing Company, Inc.

WWW.STARPUBLISHING.COM

ISBN: 0-89863-198-X

Revised printing 2006

9 8 7 6 5 4 3 09 08 07 06

Printed in Korea

Table of Contents

Preface

To the Student: *Basic Microbiology Techniques* 4th edition is intended to provide you with the necessary skills to work effectively in the exciting field of microbiology. Most of you using this manual will be encountering microbes for the first time; therefore, exercises have been selected which have proven success in class use and are typical of the skills and procedures used by biologists and microbiologists regardless of specialty. The overall objectives of these exercises are to give you repeated opportunity to practice aseptic technique, to develop isolation procedures, to provide the background principles behind the techniques, and to teach laboratory safety. Considering the microbial-laden environment around us, the development of aseptic technique should prove useful for even non-science students as well as preparing some of you for advanced work in the biology laboratory. No special background courses are necessary for students using this manual, although previous experience in chemistry and biology courses would be useful. While bacteria are the main focus of this laboratory manual, other microorganisms have been included since today's microbiologist is often required to work with and have some knowledge of these other microbes.

To the Instructor: An Instructor's Manual is available to the adopting instructor upon request. The authors and publisher would appreciate any comments on the use of this manual or its exercises.

Introduction

Basic Microbiology Techniques 4th edition has been rearranged and expanded slightly to combine some related exercises and to provide a greater selection of basic procedures. The manual is divided into eight units plus the laboratory safety section. The first four units present the very basic techniques associated with the field. The remaining four units present various applied aspects of microbiology, including the effects of environment on bacteria, identification of unknowns, health and safety of water and food, some medical aspects, and some of the newer techniques associated with mutations and immunology. The instructor may select the exercises best suited to the particular needs of the class. The manual is not intended to be used with any particular textbook and the authors have successfully used several over the years.

Each exercise presents a series of **objectives** to aid in your learning (i.e., goals to be achieved); a brief **introduction** explaining the background of the topic, which may be supplemented by the instructor, if desired; a **materials** list which, unless otherwise indicated, is on a per-student basis, assuming that each student performs the work from which the experience will come; a **procedure** describing in detail how to carry out the exercise to meet the objectives; an **observation** section that tells you what to look for; a **report form** to help you organize the results; and finally, a series of **questions** directed to the objectives of the exercise. Instructors may modify any part of a given exercise to meet class needs (e.g., the materials list is designed so that every student does the work for most of the first four units); the instructor may select other options. In this case care should be taken that each student then gets the necessary hands-on experience. Some instructors may decide to use only a few of the exercise questions rather than all. Although the questions are generally taken from material in the laboratory manual itself, a very few may require the student to seek information from a textbook. In this case most textbooks are suitable.

The initial laboratory period is usually rather hectic, assigning bench space and filling out class records. Part of this period should be directed to the laboratory safety regulations. A suggested set of regulations will be found immediately following this introduction. Most of these are standard laboratory practice; however, the instructor may wish to add others peculiar to the particular class. Unit I covers the microbial world, with a number of exercises relating to the use of the microscope. One major objective of this unit is learning to **be observant**. You should have an idea of what to expect by reading the exercise of the day **before** coming to class. As observations are made, carefully note small variations in shape, color, size, or other aspect. Although your attention will be called to most of these in class, you must be more critical in your observational acuity than ever before. One of the first things you must learn is to "see small". Bacteria, in particular, surprise most first-time observers with their minuteness. Unit II deals with media for growing microbes, isolation and counting of them, and aseptic technique. While aseptic technique is a critical part of laboratory practice, it also applies to everyday life in the prevention of foodborne illness, in the hospital, and other places where the transfer of unwanted microbes may take place. Unit III continues with techniques to aid in the observation of bacteria, while Unit IV deals with the most important part of microbial activity, metabolism (i.e., what they do and how to observe them). Unit V deals with the effect of environment on microbes and the wide range of extreme conditions under which they can live. Unit VI introduces methods used in the diagnostic laboratory to identify bacteria; the techniques rely heavily on the material of the preceding units.

Units VII and VIII detail how microbes interact with humans. Unit VII focuses on aspects involving food, water, and production and selection of laboratory mutants. Unit VIII is also an applied area but is specifically directed to medical and clinical laboratory aspects.

It is the authors' hope that you enjoy your experience with microbiology. When you finish this course you should have an expanded perspective of many things you have taken for granted in the past as well as view the world around you with new insight.

Laboratory Safety

General Requirements

1. All materials and clothes other than the laboratory manual and notebook are to be kept off and away from the bench.
2. Wear a lab coat or apron. Dyes and other materials CANNOT be removed from clothing.
3. Read each exercise carefully **BEFORE** the period so that you know what is to be done and the basic principles involved.
4. Do not begin work until a brief introduction is given by the laboratory instructor. Good laboratory technique hinges on knowing what you are to do. A good policy is to take notes in your lab manual on procedures, principles, and modifications.
5. It is extremely important that media removed from the supply containers **NOT** be returned to the containers under any circumstances. Many media are indistinguishable from one another and a mistake may ruin someone else's work.

Regulations

1. Wipe off the work bench top with the disinfectant supplied both **BEFORE** the lab starts and **AFTER** you have finished for the day. Do not assume that the previous lab user has done this. Use the same solution to disinfect your hands after washing them thoroughly with soap and water.
2. Mouth pipetting is **NOT** permitted. Use the pipet aids provided.
3. On every bench will be found containers for discarding pipets. Please keep them clean and do not put **anything** except used pipets in them.
4. Keep extraneous materials off the bench tops and put all materials in their proper places after class.
5. Use the waste baskets or other provided containers for paper and other discarded materials. **DO NOT** dispose of materials contaminated with bacteria in any container except those provided for that purpose. If in doubt, ask your instructor.
6. Discard glass in trays and test tubes in baskets. **DO NOT** mix glass and plastic. Plastics, cotton swabs, and other disposables should be discarded in plastic autoclave bags. If appropriate trays, baskets, or bags are not available, see your instructor. **DO NOT** leave tubes or plates simply lying about where they can fall over.

7. Remove all wax or indelible pen marks from glassware with xylene or other solvent provided for that purpose.
8. Because the organisms used in this class are potentially (but rarely) pathogenic, aseptic technique is of paramount importance. Develop good habits at the beginning. Keep fingers, hands, and objects away from your mouth. Labels that require licking are **NOT** permitted. **Eating**, **drinking** (soft drinks or other fluids), and **smoking** are prohibited in the laboratory for your own safety.
9. Children are not permitted in the laboratory under any circumstances.
10. Report all accidents such as cuts, burns, or spilled cultures to your instructor *immediately*. Students with long hair should be especially cautious around the Bunsen burners.
11. Microscope lenses should be wiped clean of oil and then put away with the low power objective down. If these are not in the proper condition when you take the microscope for use, let the instructor know immediately.

UNIT I

Microscopy and the Microbial World

Antonie van Leeuwenhoek's first report in 1677 on his observation of microbes has led to the discovery of a vast world of living things intimately involved in our everyday life—medicine, foods, soil, and chemistry to name a few. Since magnification was required to visit this world, the microscope has assumed a central role in microbiology and remains one of its most important tools even today. It is difficult to define a microbe, but it is commonly referred to as a living single cell requiring a microscope to see the entire organism. Some multicellular animals (e.g., helminths) are included in microbiology, because the microscope is the major tool to visualize them.

The world of microbes is vast, made up of prokaryotic and eukaryotic organisms. A wide range of sizes exists from viruses (which are not really cells) to fairly large structures seen with the naked eye. Most are single cells, but a few of the larger forms are multicellular. The fascinating array of these organisms includ es free-living forms, saprophytes, parasites, pathogens, and symbionts, all playing major roles in the biosphere. A few of the major groups considered to be microbes include:

Bacteria. These are the smallest of the cells. All are prokaryotic and found universally in the biosphere. Many are important as decomposers and nutrient cyclers. A few are pathogenic.

Cyanobacteria. These photoautotrophic prokaryotes are important oxygen producers and nitrogen fixers. A few are symbionts and some form toxins.

Algae. These eukaryotic photoautotrophic oxygen producers are rivaled in biomass only by the plants. Some are symbionts and a few form toxins.

Protozoa. These eukaryotic unicellular organisms feed on bacteria and other materials. A few are pathogens. Some are symbionts and some have symbionts of their own, including algae and bacteria.

Fungi. These eukaryotic decomposers are found universally in soils and foods and are important antibiotic producers. A few are pathogens.

Helminths. These multicellular eukaryotes are usually classified with the animals. Many cause important human and animal diseases (e.g., schistosomiasis).

Viruses. Although not cells, these are all parasites of living cells, including all of the microbes listed above. They use the host cell biochemical machinery for reproduction.

This unit is intended to introduce students to a minimum of optics, the care and proper focusing of a microscope, while observing a few representatives of this most diverse and interesting group of living things, the microbes.

1
Care and Use of the Microscope

Objectives

The student will be able to:

1. identify the parts of the microscope and their respective functions.
2. define the terms total magnification, resolving power, and working distance.
3. properly handle and care for the microscope.
4. focus the microscope properly (in conjunction with Exercise 2).

Microbiology holds as one of its principle unifying bases the need of a microscope to observe the organisms involved. There are many types, such as bright field, phase contrast, interference, and electron microscopes, differing in manner of construction and details of operation. However, certain principles underlie all microscopes.

MAGNIFICATION results from the use of one or more lenses. In a compound microscope, the objective lens nearest the object magnifies, producing a real image. The eyepiece or ocular lens magnifies the real image, producing a virtual image seen by the eye. The magnification of a lens is usually expressed in diameters of apparent increase in size or power (e.g., 10X is a ten times increase in size or 10 power). Microscopes generally consist of at least three objective lenses: a low-power objective of about 10X, a high-power objective of about 40X, and an oil immersion lens of about 100X (Figure 1-1). The eyepiece lens is usually about 10X. The actual magnification of the lenses depends on the type of microscope and the manufacturer. The total magnification of the microscope is the product of the magnification of the objective and the magnification of the eyepiece. Some microscopes have an additional lens or lenses between the objective and eyepiece and this magnification must also be multiplied by that of the objective and the eyepiece.

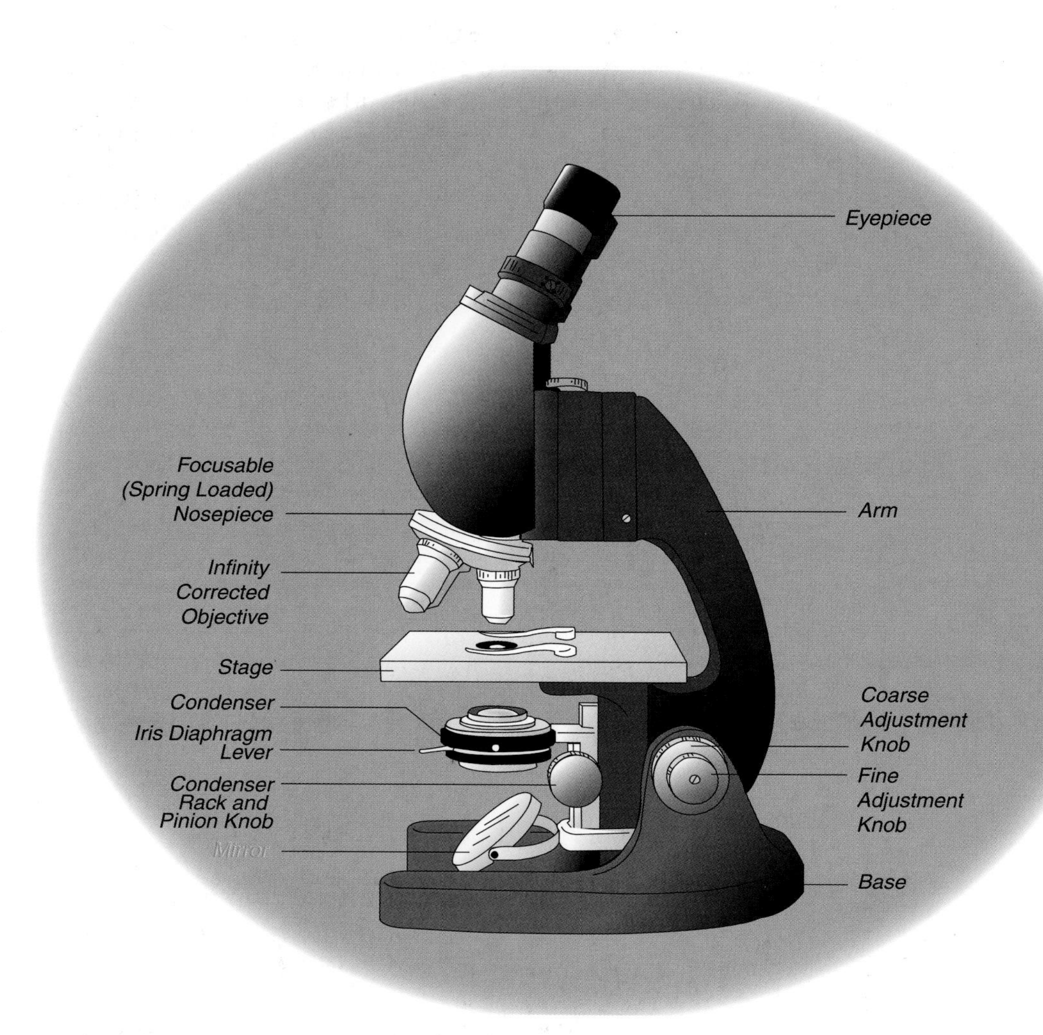

Figure 1-1

Compound light microscope

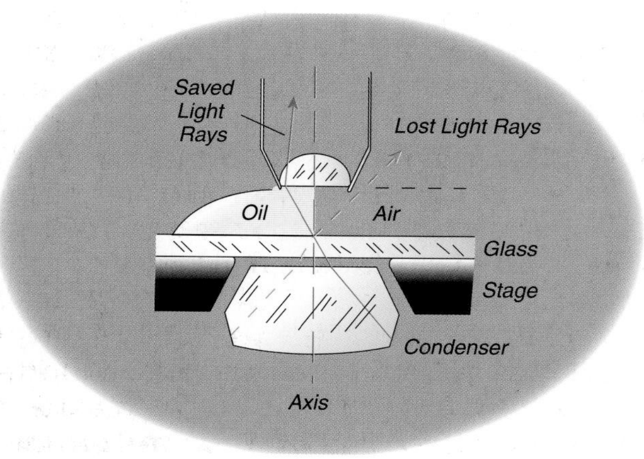

Figure 1-2

Effect of oil on light gathering in an objective lens

RESOLVING POWER is the ability of a lens or a microscope to distinguish between two closely adjacent points. The theoretical resolving power of a microscope is rarely achieved due to lens aberrations and the fact that extraordinary performance on the microscope by the user is required. The resolving power is a function of the wavelength of light used and the numerical aperture of the lens. Numerical aperture (NA) refers to the ability of a lens to gather light and is expressed by

$$\text{NA} = \text{i} \sin \theta$$

where θ is one-half the angle of the cone of light entering the objective and i is the index of refraction of the medium between the object and the objective. The index of refraction for air is 1.0. In order to increase the light-gathering ability of the lens, something with an index of refraction greater than 1.0 must be placed between the object and the lens. Immersion oil serves this purpose, having an index of refraction nearly identical with the glass of the lens, about 1.56 (Figure 1-2). The condenser also acts as a lens and thus has a numerical aperture as well.

Resolving power can be calculated as follows:

$$d = \frac{\lambda}{\text{NA of objective} + \text{NA of condenser}}$$

where d = diameter (μm) of the smallest observable object, λ = wavelength (μm) of light, and NA is numerical aperture (unitless). If the condenser NA is unknown, two times the NA of the objective is used (i.e., 2 x NA of objective).

The resolution should be better as wavelength decreases but is limited by several factors. The human eye is most sensitive to green light and "sees" best at about 500 nm. Green or blue green filters are often used to aid in this. Photography works best at ultraviolet wavelengths and special lenses are used. For optimum resolving power, in addition to an expert microscopist, oil must be interposed between the condenser and the bottom of the slide. This is not usually done in the teaching laboratory.

WORKING DISTANCE is the space between the objective and the specimen when the latter is in focus (Figure 2-4). The higher the magnification, the smaller or the shorter the working distance, and generally speaking, the greater the amount of light required to see the specimen. The diaphragm of the condenser should be almost completely open when using the oil-immersion lens (Figure 2-4). A microscope is parfocal when one objective can be exchanged for another in position over the object, requiring only minor adjustment in focus.

Five types of light microscopes are widely used in microbiology today: bright field, dark field, phase contrast, interference, and epiluminescence. The bright field microscope is the most widely used, usually with stained preparations and is illustrated in Figure 1-1. Dark field observation is achieved by attaching an annular ring below the condenser on a bright field microscope (or by using a special condenser). The only light entering the objective is that striking the object on the slide. If there is no object, no light enters the objective. This type of microscopy is widely used to examine specimens for spirochetes, which are difficult to see with the bright field microscope.

Phase contrast microscopes are used to view specimens by increasing the contrast between the object (or subcellular components) and the background. Living organisms do not absorb light but delay it about 1/4 wavelength. Light waves delayed by 1/2 wavelength cancel undelayed light from the same light source. This cancellation creates a darkened place in the image that the eye sees. This is accomplished by the use of an annular diaphragm below the condenser and a phase plate cemented in the objective. Without a specimen on the stage, all the light is focused to pass through the thin part of the phase plate. When a specimen is present, image-forming light is directed through the thick part of the phase plate and the remaining light through the thin part. The details of the specimen delay the wavelength by 1/4 and the thick part by another 1/4. When the image-forming light mixes with light from the thin part of the phase ring, it creates a darker image, allowing transparent detail to be seen. Unfortunately, a halo is also created around the outside of the object image which sometimes interferes with seeing the object.

Interference contrast is similar to phase contrast except that a polarizer and a Wollaston prism are used below the condenser to create two light paths, one through the specimen and the other through the background, which are then recombined to interfere with one another, thus enhancing contrast without the halos common in phase contrast.

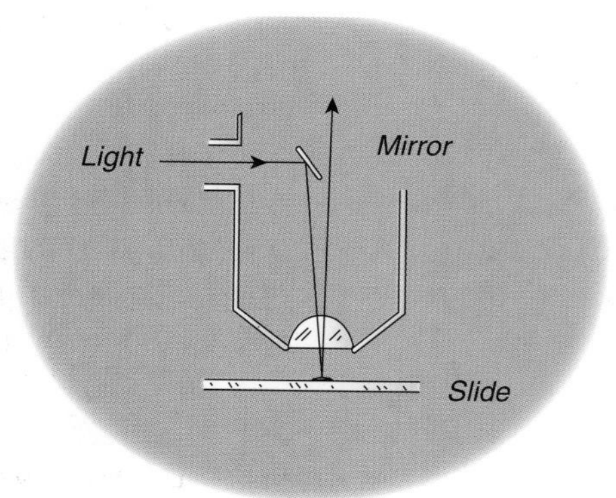

Figure 1-3

Epiluminescence microscope

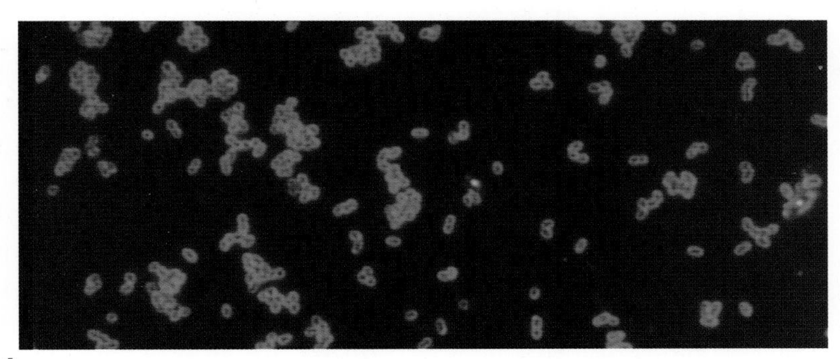

Figure 1-4

Bacteria as seen by fluorescence microscopy

Epiluminescence microscopy is a form of bright field microscopy in which light passes down through the objective, strikes the object, and returns upward through the objective again. This type of system is widely used with fluorescence observation in which ultraviolet light causes an object to fluoresce in the visible range with the light returning to the eyes.

Each lens has letters or numbers giving some information about the lens. The letters often refer to the use of the lens. Ph means phase, Fl means fluorescence, Plan means a flat field of view, Achrom means color-corrected (achromatic), etc. The number followed by X is the magnification. Sometimes there is no X but a / followed by the numerical aperture, a number between 0 and 1.6. Some older lenses include the focal length, a number followed by mm; newer lenses, however, do not include this feature.

Materials

1. Microscope
2. Exercise 2

Procedure and Observations

Period 1

Note: This exercise is intended to be accompanied by Exercise 2 for focusing instruction.

1. A microscope will be assigned to you for use in class. You should use only the assigned microscope and its proper care is your responsibility. If you find it improperly stored, with oil on the lens, or a lamp burned out, report it immediately to your instructor. Do not use any other microscope without your instructor's approval.
2. Grasp the microscope by the arm (Figure 1-5) with one hand. Place the other hand under the base and return to your bench. Carry the microscope upright, being very careful not to tip the body tube allowing the eyepiece to fall out. Place the micro-

scope on the bench and identify each part using Figure 1-1 as a guide. Study the introduction to this exercise and identify the symbols on the lenses. Since variations are common among manufacturers, your instructor may provide additional information about the symbols or the microscope.

3. When returning the microscope to its storage place, make sure all oil has been removed from the lenses. Turn the low-power objective down. Make sure the dust cover is in place, if required.
4. After completing your study of the microscope, answer the questions and submit them to your instructor.
5. Proceed to Exercise 2 for focusing of the microscope. These exercises can be done independently if desired.

Precautions

1. Clean the lenses only with lens paper or tissue paper (Kleenex) provided by your instructor. Clean the eyepiece lens occasionally, as eyelashes leave an oily film.
2. Always remove the slide and clean the lenses when finished.
3. Clean the stage and condenser if oily. Keep them clean.
4. Use two hands to carry the microscope. One hand grasps the microscope arm, the other hand is under the base. Do not tilt the microscope when carrying it.
5. Do not force parts. Call your instructor if a problem arises.
6. Always raise the objective (or lower the stage) when focusing so that the lens and slide are moving apart.
7. Store your microscope with the low-power objective down and no oil on lenses or stage.

Figure 1-5

Carrying the microscope

2 Bacteria and Cyanobacteria (Focusing)

Objectives

The student will be able to:

1. properly focus the microscope (in conjunction with Exercise 1).
2. name some of the morphological forms of bacteria and cyanobacteria.
3. recognize heterocysts and akinetes in the cyanobacteria.
4. name the genera of five morphological forms of bacteria.
5. name the genera of four morphological forms of cyanobacteria.

The **eubacteria** (**true bacteria**) and **cyanobacteria** represent the prokaryotic cell types among the microbes. The bacteria are the smallest cell types, averaging less than one micrometer (μm) in diameter. The cyanobacteria average slightly larger. Morphologically the bacteria are fairly limited in types (Figures 2-1 and 2-2), and classification is only partially based on shape. The cyanobacteria are more varied, with a number of specialized cells such as heterocysts and akinetes (Figure 2-3) found in some. Extensive branching occurs in morphologically more advanced forms of both groups. The cyanobacteria also contain chlorophyll *a* (only) and when combined with accessory pigments, often colors the cells a green, blue-green, or brown. The color may be seen in cells of larger forms under the microscope. Chlorophyll *a* does not occur in the bacteria and is replaced by bacteriochlorophylls of several varieties. Although bacterial colonies are often colored, pigments cannot be seen in individual cells under the microscope.

Some cyanobacteria exhibit motility of a gliding type (also found in a few bacterial groups), otherwise they are nonmotile. Gliding occurs only when a cell or filament is in contact with a surface (e.g., a coverslip or slide). **Akinetes** are special reproductive cells formed by some cyanobacteria. **Heterocysts** are rounded specialized cells in which nitrogen fixation takes place, often appearing very refractile (bright and shiny) under the microscope. These structures are usually observed in filamentous forms under low available nitrogen conditions. The location of these in the filament is useful in taxonomy of the filamentous forms.

This exercise is intended to show only a few of the many morphological forms of bacteria and cyanobacteria occurring in nature and a eukaryotic yeast for comparison. It also provides an opportunity to practice use of the microscope as described in Exercise 1. It should be noted that few, if any, intracellular structures can be seen in either group.

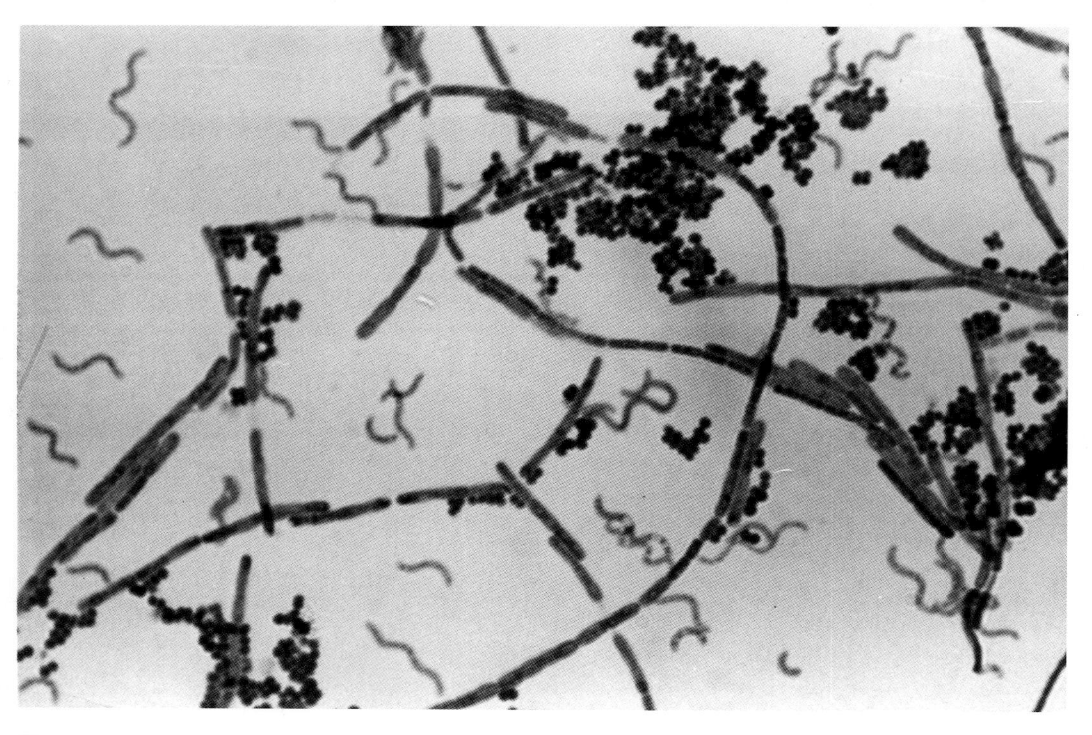

Figure 2-1

Oil immersion view of bacterial rods, cocci, and spirilla

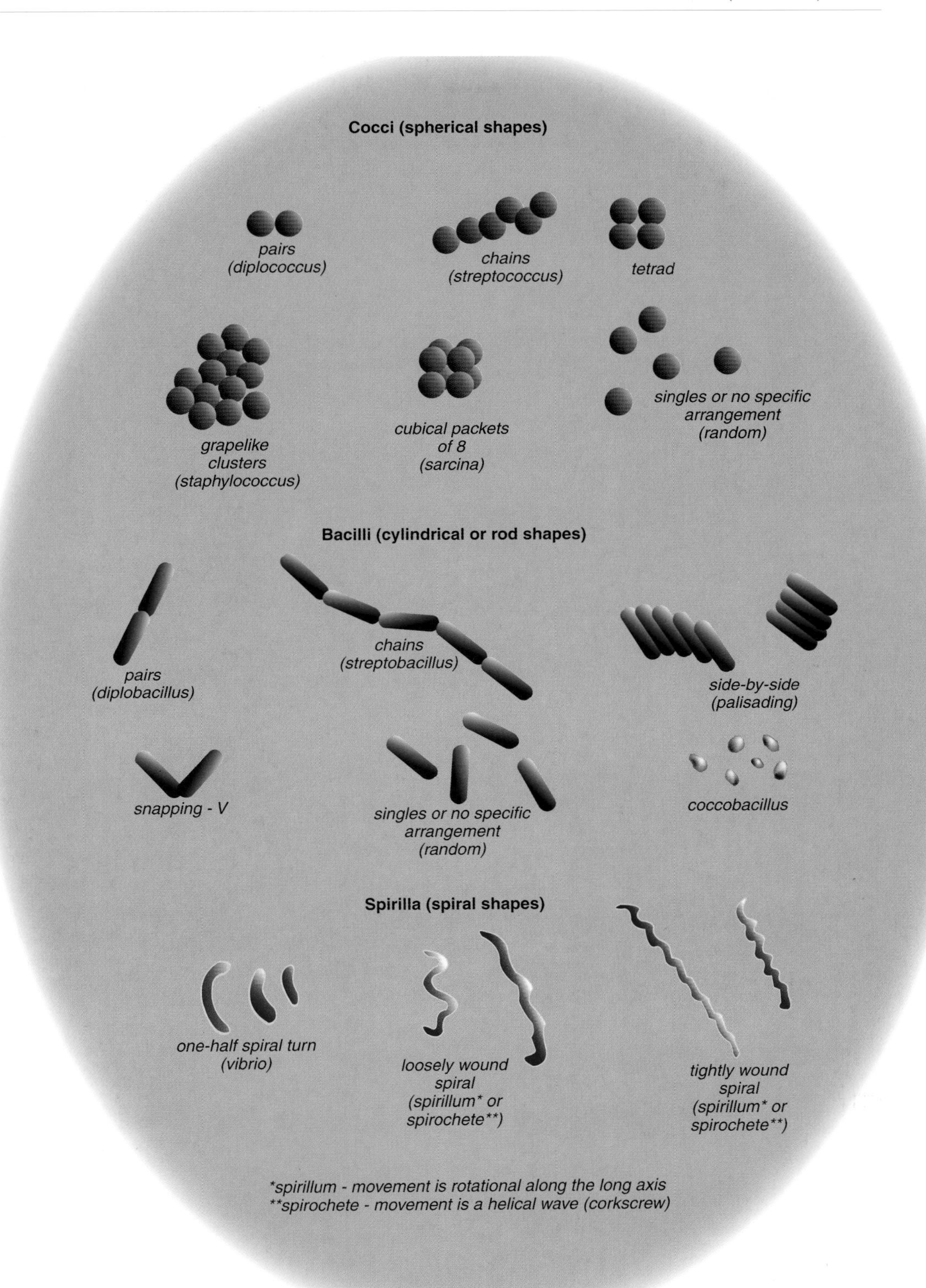

Figure 2-2

Bacterial shapes and arrangements (morphology)
These terms are not used for cyanobacteria.

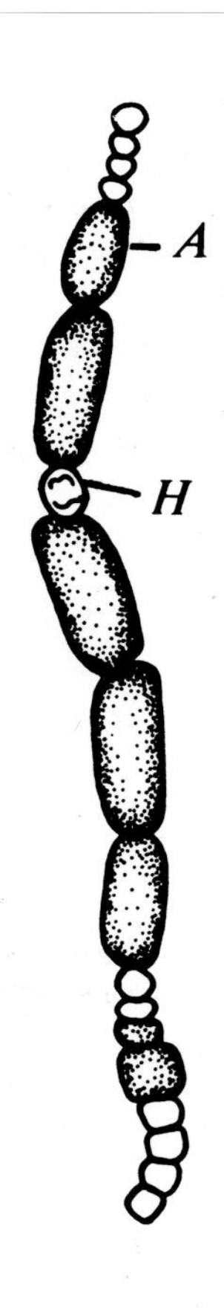

Figure 2-3

Cyanobacteria. A is akinete; H is heterocyst. (Adapted from G. E. Fogg et al. "Blue-Green Algae", Academic Press, NY 1973.)

Materials

1. Prepared slides of bacteria, cyanobacteria, and a yeast

Bacteria	Cyanobacteria	Yeast
Bacillus subtilis	*Anabaena*	*Saccharomyces cerevisiae*
Proteus vulgaris	*Oscillatoria*	
Staphylococcus aureus	*Spirulina*	
Streptococcus pyogenes	*Microcystis*	
Spirillum sp.		

2. Microscopes
3. Immersion oil
4. Lens paper

Procedure

Period 1

1. Place the prepared slide on the stage, specimen side **up**.
2. **Center** the stained specimen over the light hole in the stage as carefully as possible.
3. **Lower the condenser** about 3/4 of the way to the stop and **close the diaphragm** nearly completely (Figure 2-4). **Lower the low-power objective** over the specimen until it cannot be lowered further.
4. Looking through the eyepiece, slowly **raise** the objective (or lower the stage, with some microscopes) using the coarse adjustment until some color is in view indicating stained cells.
5. Then focus carefully **using the fine adjustment *only*.** Adjust the condenser and diaphragm (Figure 2-4), if necessary. After centering the specimen (note that no details can be seen, only color), move your eyes to the level of the stage and, while looking from the side, rotate the **high-power** objective over the specimen until it clicks into place. **Raise the condenser** until it is about 2 cm below the stage. **Open the diaphragm** 1/2 to 3/4 of the way (Figure 2-4).
6. Again, look through the microscope focusing with the **fine adjustment *only*.** Adjust the condenser level and diaphragm as needed (Figure 2-4). Note that, at least with the bacteria, only a little more detail can be seen. Center very carefully. Rotate the oil immersion lens about half way, and place a drop of immersion

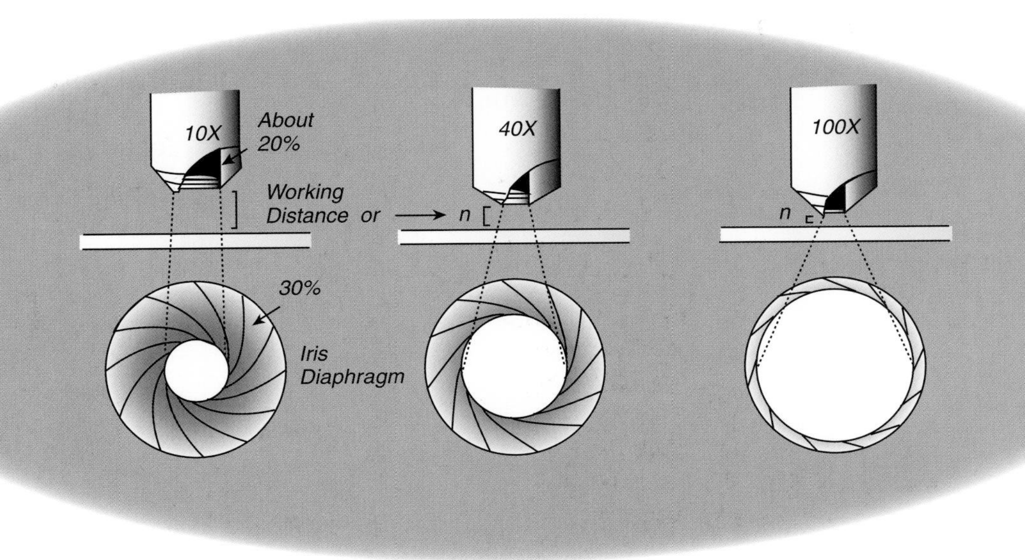

Figure 2-4

Approximate diaphragm openings and working distance

oil on the specimen over the center of the light path. Then with your eyes again at the level of the stage, rotate the oil lens into the immersion oil. If there is ***ANY*** resistance, stop immediately and determine why. Most lenses are parfocal, so it may mean a problem has arisen. Ask your instructor for help if the cause is not readily determined. **Raise the condenser to its full upright position and open the diaphragm completely** (Figure 2-4).

7. Look through the microscope focusing with the **fine adjustment *only***. The slide may have to be moved a little to bring cells into the field of view. Adjust the level of the condenser and the diaphragm opening (Figure 2-4), if necessary. Focus carefully and make your observations.

Observations

Period 1

In making drawings, use ink and/or colored pencils, preferably the latter.

1. **Bacteria.** You will find these very tiny. Draw a few representative cells of each organism in a circle on the laboratory report form. Note the color of the cells in the space provided. Make drawings of cells reasonably large. Do not simply make a pencil line to represent a bacillus. From Figure 2-2 give the morphological name for each bacterium.
2. **Cyanobacteria.** Draw a few representative cells from each preparation and especially look for heterocysts and akinetes. Note that the morphological names used for the bacteria (Figure 2-2) are NOT used to describe cyanobacteria. Use terms such as filamentous, spherical, and branching instead.

   Cyanobacteria are best seen as live preparations. If they are available, your instructor will give special instructions for preparation and/or observation.

3 Fungi

Objectives

The student will be able to:

1. recognize the microscopic appearance of fungi and their asexual spores.
2. distinguish septate from nonseptate fungi.
3. differentiate yeasts from molds.
4. prepare a slide culture of fungi (if done).
5. make a wet mount for microscopic observation (if done).

The fungi are a complex group of **eukaryotic** microbes ranging from the unicellular yeasts to the extensively mycelial molds. Some are able to form visible sexual fruiting structures variously called mushrooms or toadstools. The fungi are placed in the Kingdom Fungi in the five-kingdom system of taxonomy. The fungi are important as contaminants in the laboratory, as a food, in creating new foods, in producing antibiotics and industrial chemicals, as pathogens, and in their own right. Some of the more common pathogens are described in *Control of Communicable Diseases in Man*, A.S. Benensen, ed. 16th edition, 1995. American Public Health Association, Washington, DC.

The growth form of most fungi is a tubular cell called a **hypha** (pl. hyphae) which grows from the tip and may form many branches. This many-branched hypha (**mycelium**) can become large enough to be seen with the naked eye, when it is often referred to as a **mold** (Figure 3-1). The term "mold" is not rigorously defined. One morphological characteristic dividing the fungi is whether or not the hyphae possess **septa** or cross walls (sing., **septum**) (Figure 3-2). The lower fungi (previously called the **Phycomycetes**) have no septa and the cytoplasm and nuclei migrate freely through the hyphae. This condition of free movement is called **coenocytic**. The other group has discernible septa across the hypha at intervals, but these cross walls are not complete. A hole is present in the center

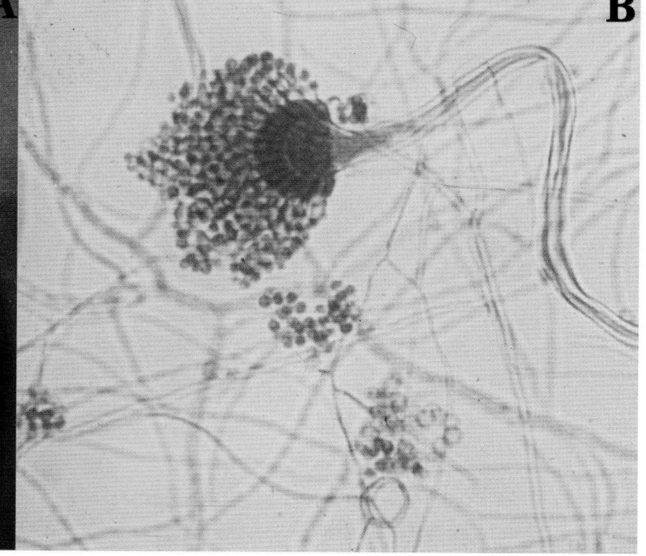

Figure 3-1

Typical molds. (A) Plate showing colonies; (B) *Aspergillus* sp.

of each septum, allowing the coenocytic condition.

Fungi are classified on the basis of the type of sexual reproduction. If sexual spores are formed in a **basidium**, they are classified as **Basidiomycetes**. Typical representatives are the mushrooms and toadstools. If sexual spores are formed in a structure called an **ascus**, they are classified as **Ascomycetes**. Typical representatives are *Aspergillus* and *Penicillium*. Enclosed sexual spores of other types are classified among the lower fungi. Examples are *Rhizopus* (bread mold) and *Saprolegnia* (water mold). If no sexual stage has been observed, the organism is placed in the **Deuteromycetes** (or Fungi Imperfecti).

Although the hyphal form is common among the fungi, some exist only as single cells or at most as a pseudohypha a few cells long, possibly even exhibiting a little branching. These are known as the **yeasts**. Perhaps the best known is *Saccharomyces cerevisiae* (Ascomycetes), the bread, beer, and wine yeast (Figure 3-4). Many fungi have two morphological forms, a yeast-form under some conditions and a mycelium-form under other conditions. Such fungi are called **dimorphic** and include some important human and animal pathogens.

Asexual spores are formed by most fungi and vary widely in morphology and mode of formation. Asexual spores of most lower fungi are enclosed, while the higher fungi form free asexual spores (Figure 3-3). Many aquatic lower fungal spores, both sexual and asexual, are motile with flagella indicating a close affinity with the flagellates of the protists. The asexual spores are often brightly pigmented, giving the colony a characteristic color (e.g., red, green, blue, black, brown, etc.). A good example is the blue spores of *Penicillium roqueforti* found in blue or Roquefort cheese. As with the other organisms in this section, specialized courses in mycology are available.

This exercise introduces the student to the fungi and several methods for observing them. It consists of three independent parts: one using prepared slides, a second illustrating a method of growing filamentous forms, and the third, a wet mount technique for observing yeasts. The instructor may choose to do one, two, or all of the independent parts below.

Part I. Observation of Prepared Slides

Materials

1. Prepared slides of a selection of the following fungi:
    a. *Penicillium*, *Aspergillus*, and *Rhizopus* on same slide
    b. *Aspergillus*
    c. *Penicillium*
    d. *Rhizopus*
    e. *Rhizopus* conjugation
    f. Budding yeast
    g. *Saprolegnia*
    h. *Candida albicans*
    i. Powdery mildew of willow

Procedure and Observations

Period 1

1. Place the prepared slide on the microscope stage and focus with the low-power objective as in Exercise 2.
2. Carefully rotate the high-power objective into place and focus, adjusting the condenser and diaphragm as necessary.
3. Most observations will be made with the low- and high-power objectives. Make drawings on the report form of hyphae, branches, fruiting structures, spores, rhizoids, stolons, and other structures, as present, for each fungus provided, using Figures 3-2, 3-3, and 3-4 as guides.

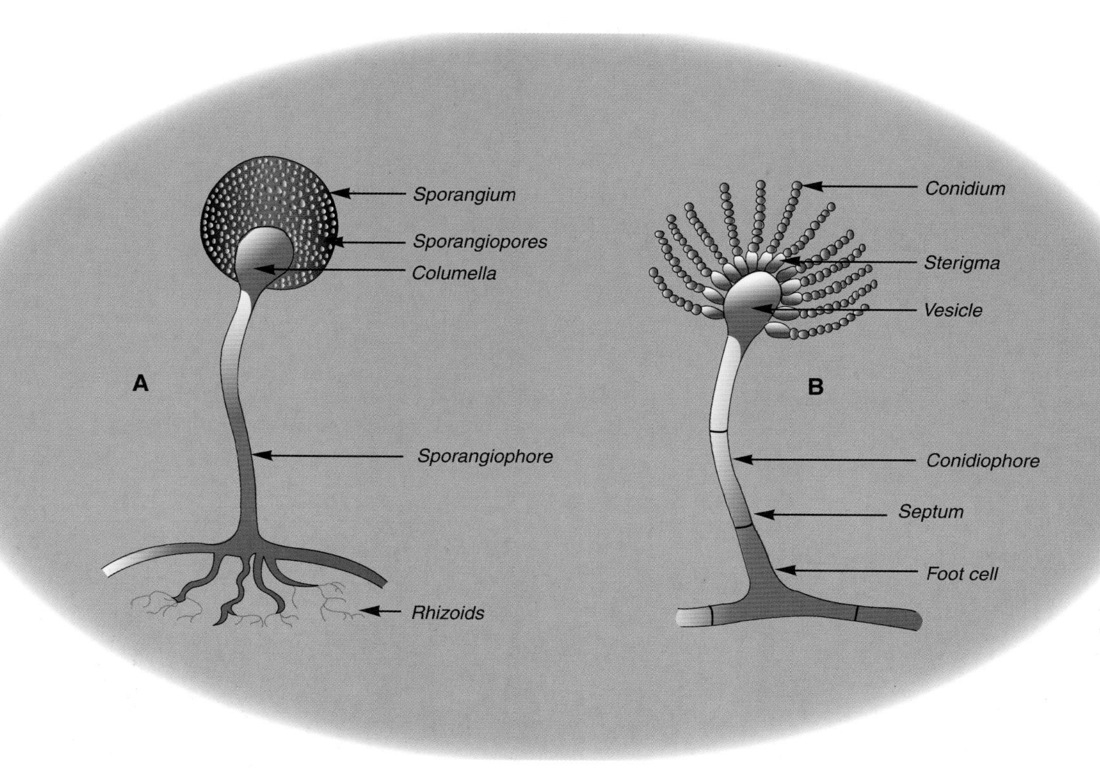

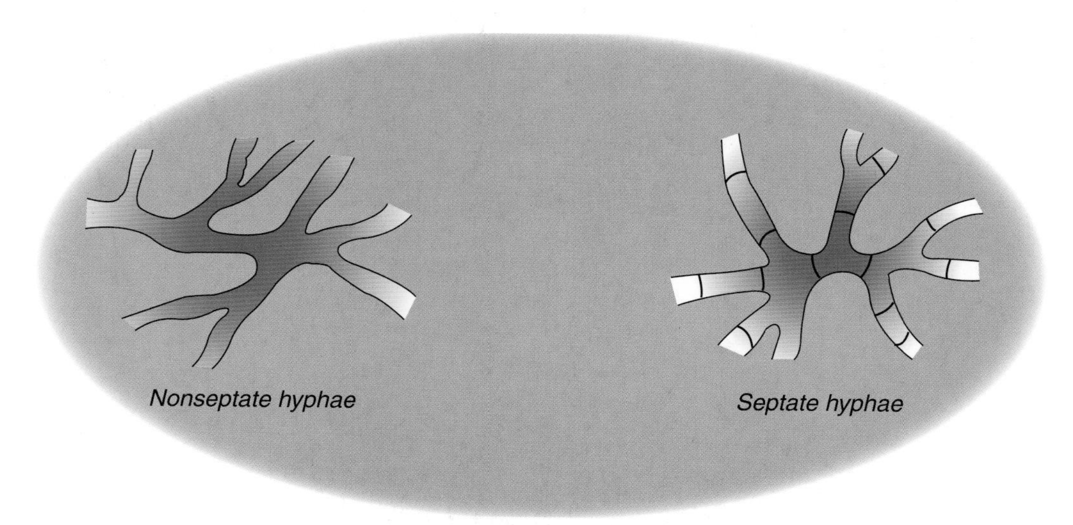

Figure 3-2

Asexual reproductive structure of *Rhizopus* (A) and *Aspergillus* (B), including respective septate and nonseptate hyphae

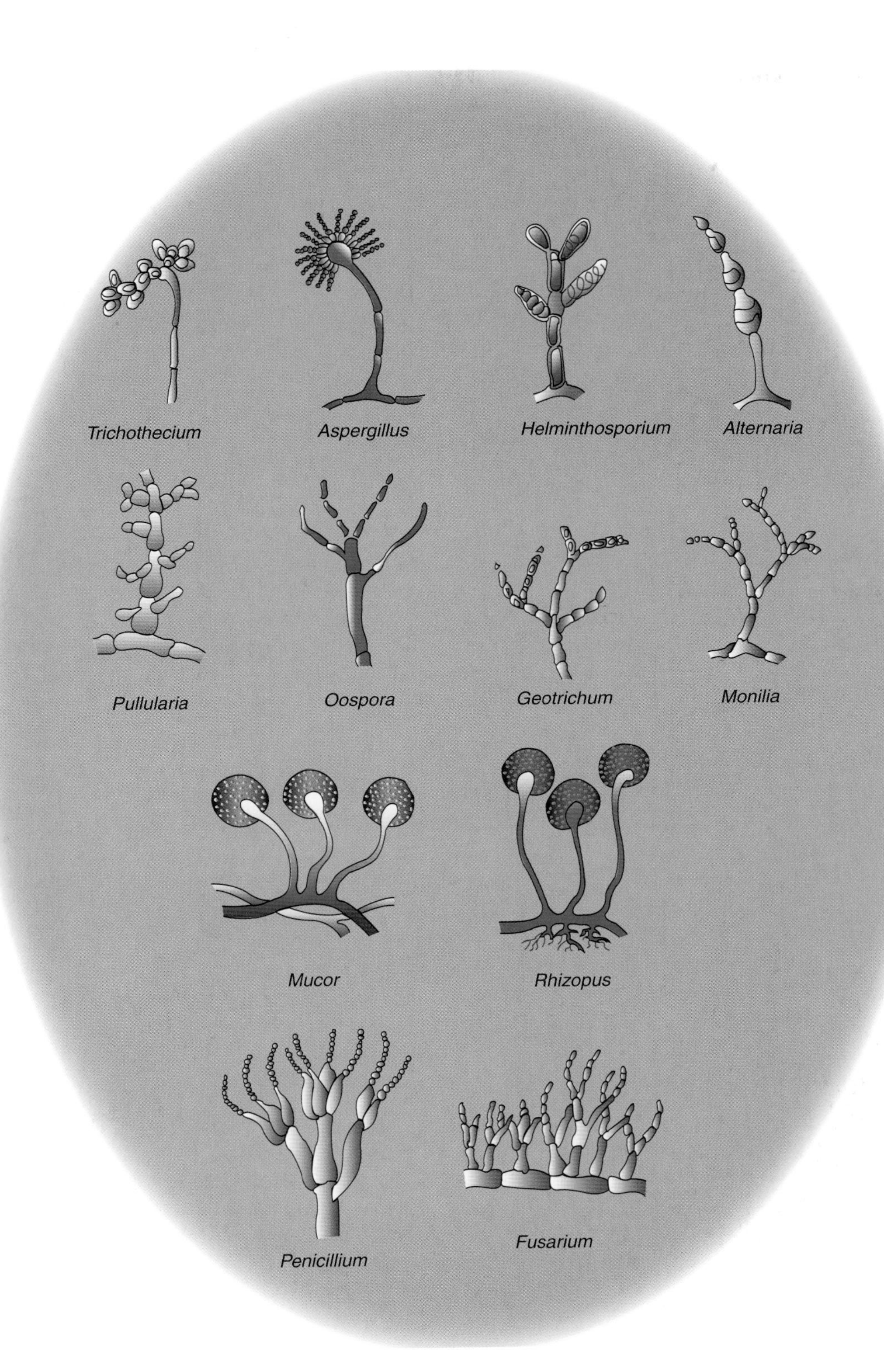

Figure 3-3

Asexual reproductive structures of commonly found fungi

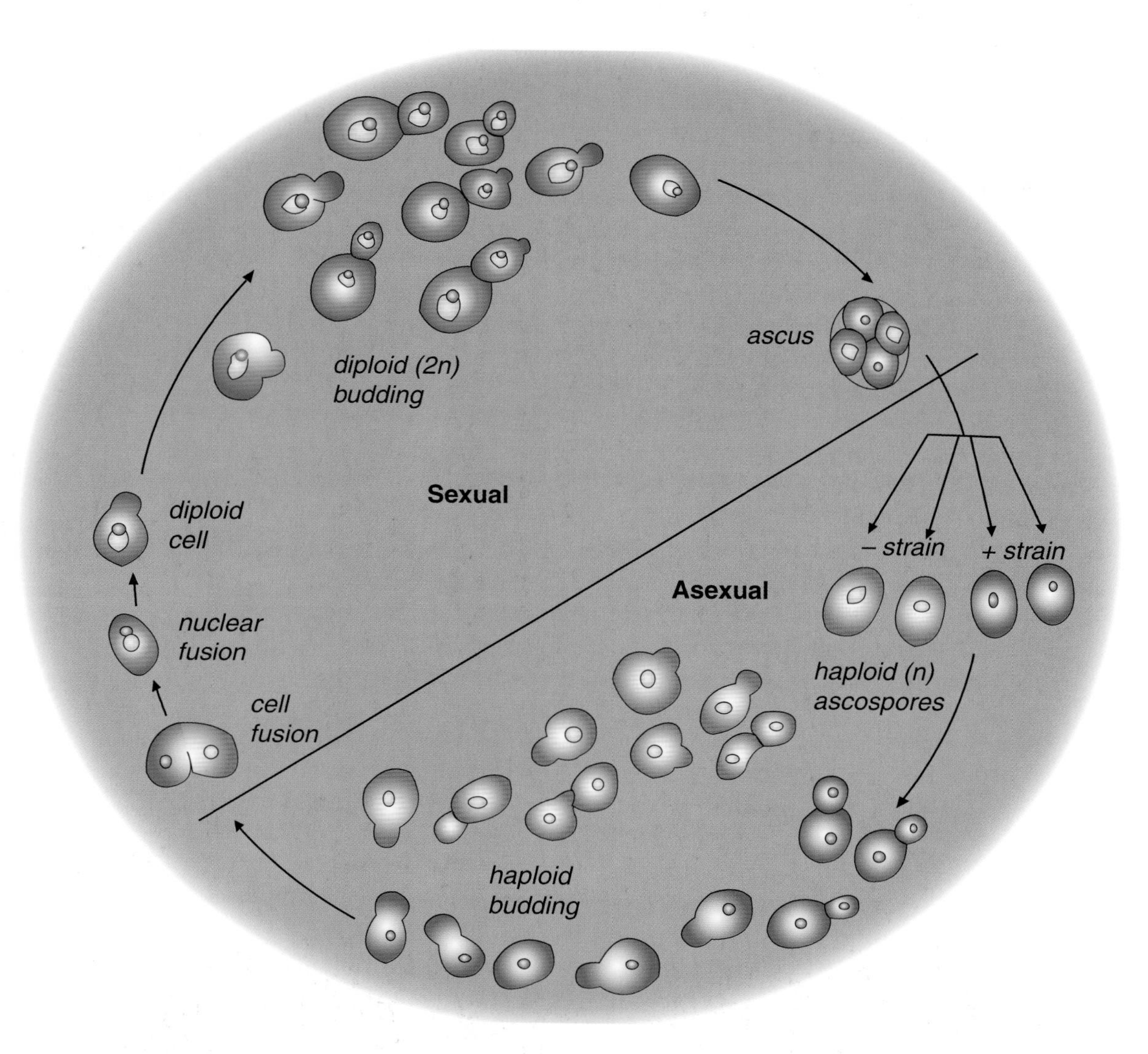

Figure 3-4

Diagram of life cycle of *Saccharomyces cerevisiae*

Part II. Growth of Fungi in a Moist Chamber

Materials

1. 1 sterile glass Petri dish with fitted filter paper, glass slide, and coverslip
2. A small amount of Sabouraud's dextrose (or maltose) agar
3. 2 sterile Pasteur pipets with bulb
4. Vaspar, sterile, melted
5. 99 ml dilution blank of sterile distilled water
6. Forceps
7. Mold cultures: *Penicillium*, *Aspergillus*, and *Rhizopus* or wild cultures

Procedure

Period 1

1. Obtain one sterile glass Petri dish with coverslip, slide, and filter paper in it.
2. Using a sterile Pasteur pipet, add 1 to 2 drops of melted Sabouraud's dextrose (or maltose) agar on the center of the glass slide and let it solidify.

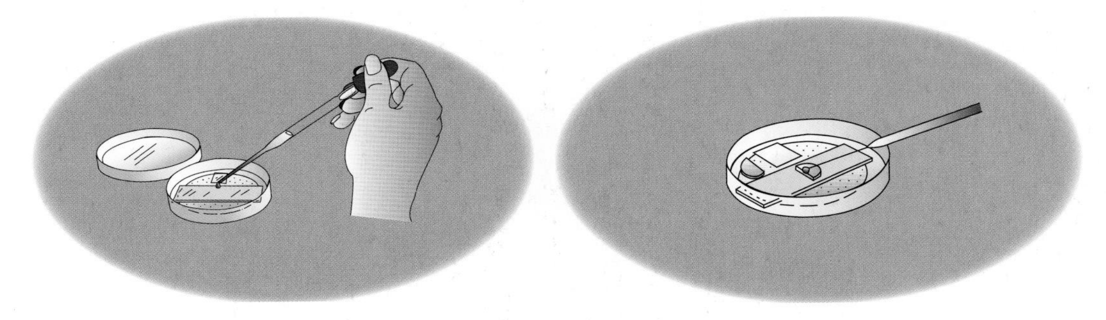

3. Using a flamed and cool loop, cut the solid agar drop in half. Push one of the halves off the slide onto the filter paper.
4. Inoculate the cut face of the drop on the slide with one of the fungus cultures or unknowns provided.

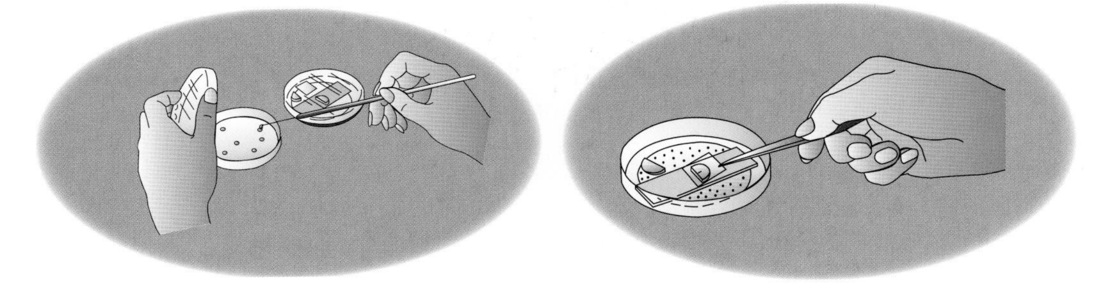

5. Sterilize the forceps in the flame and allow to cool. Pick up the coverslip with the forceps and place it on top of the agar drop. Make sure the surface is parallel with the slide.

6. With a sterile Pasteur pipet, seal the coverslip-slide space around three sides, leaving the side opposite the inoculated cut side open to the air. Avoid getting vaspar on the top of the coverslip.

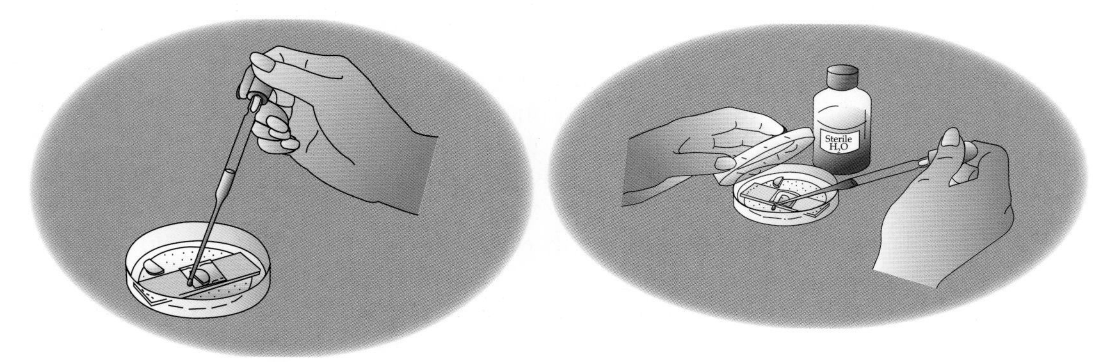

7. Add just enough sterile water to the filter paper to completely wet it. Check the dish each day and keep the paper moistened until sufficient growth is present.
8. Incubate the covered plate at 20°–30°C until desired growth is obtained. About one week is usually sufficient.

Observations

Period 2

1. Place the incubated slide on the microscope stage and examine the growth with the 10X objective.
2. Focus the microscope on the hyphae and spores just below the surface of the coverslip. Carefully rotate the 40X objective into position but be careful not to crush the coverslip with the objective. DO NOT attempt to use the oil immersion lens.
3. Make drawings of hyphae and spores for each fungus on the report form, looking carefully at hyphae, septa, branching, fruiting bodies, spores, rhizoids, stolons, and other structures. Use Figure 3-3 as a guide.

Part III. Wet Mount and Yeast Morphology

Materials

1. *Saccharomyces cerevisiae* and other yeast cultures on Sabouraud's dextrose (or maltose) agar or acetate agar
2. Coverslips
3. Slides
4. Pasteur pipets
5. Loeffler's methylene blue stain
6. Vaseline
7. Prepared slide of ascus with ascospores

Procedure and Observations

Period 1

1. Prepare a wet mount from the yeast culture provided by ringing a coverslip with a thin bead of Vaseline. Place a small blob of Vaseline on the rear of the palm of your hand opposite the thumb. Touch one edge of a coverslip to the Vaseline and carefully scrape the edge along the rear of the palm until a small ridge of Vaseline forms along the edge. Repeat for each edge of the coverslip until all four sides have a small bead. The bead must be thin and uniform. The Vaseline may also be added to the edge by drawing a Vaseline-coated toothpick along each edge. This method often leads to too thick a bead, however.
2. Place a small drop of distilled water in the center of the cover-

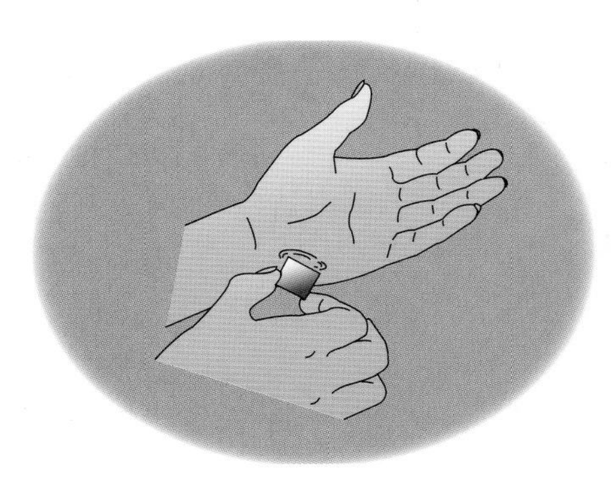

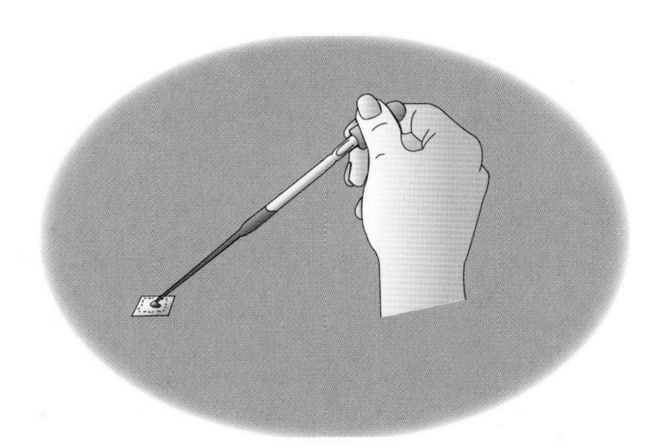

slip using a Pasteur pipet. The drop must be small.

3. Add a very small amount of Loeffler's methylene blue stain to the drop with an inoculating loop.
4. With a sterile inoculating loop, emulsify a small amount of

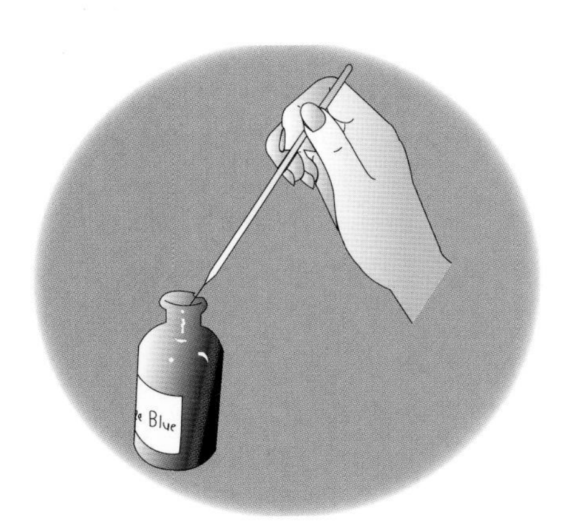

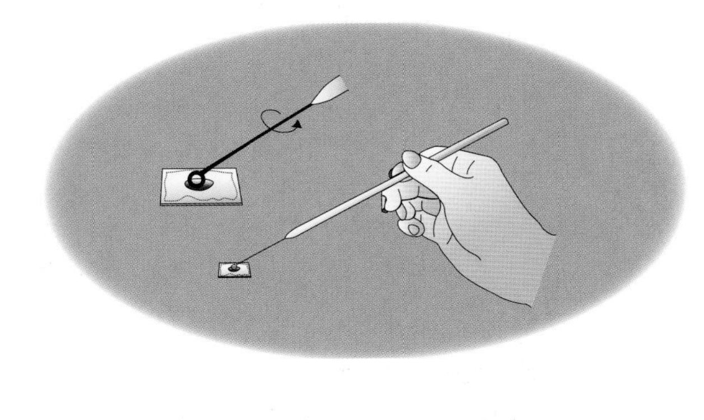

growth from a slant culture in the drop until it appears slightly murky but not milky.

5. Center a slide over the coverslip and press down gently to seal the edges.
6. Invert the slide with the coverslip up and examine it under the microscope. Make drawings using the oil immersion lens. Be

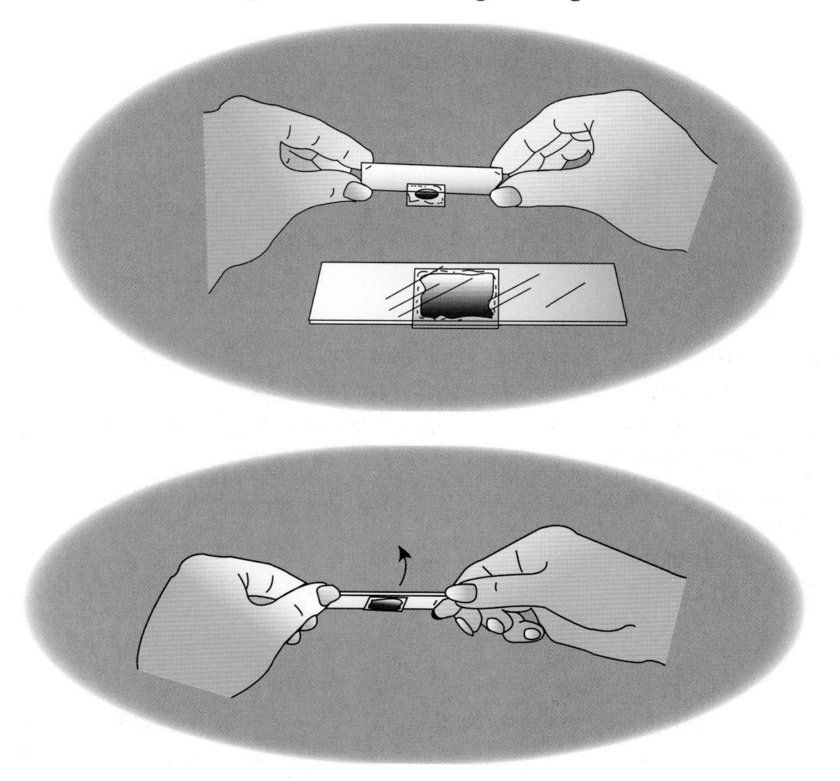

careful that the oil lens does not contact the coverslip when rotating it.

7. Look for the various stages of the life cycle of *S. cerevisiae.* Use Figure 3-4 as a guide.
8. If prepared slides are available, make observations and drawings of these.

4
Protozoa and Algae

Objectives

The student will be able to:

1. describe the meaning of the term "Protist".
2. explain the apparent overlap of classification of the motile photosynthetic protists.
3. describe the morphology of the major groups of protozoa and algae.
4. name at least four human diseases caused by protozoa and the groups to which they belong.
5. name a genus representative of the euglenids, diatoms, and green algae.

The protozoa and algae are **eukaryotic** microbes classified in a single group (except some green algae) called the **Protists** or **Protoctista**. The Fungi, Animalia, and Plantae make up the rest of the eukaryotic organisms. The close relationship between the protozoa and the algae is evident in the dual classification of the motile (and some nonmotile) photosynthetic (and some not) single-celled organisms. In many cases two genus names are given to a single species, one in the protozoa and one in the algae classification. Fortunately, the species name is usually the same. The more "advanced" algae are clearly not protozoa and vice versa, and there is no problem of dual taxonomy.

The algae are photosynthetic eukaryotes that possess a membrane-bound nucleus and chloroplast (thus excluding the cyanobacteria previously called the blue-green algae). They are a diverse group of organisms ranging from unicells (motile or nonmotile) to very large multicellular plant-like organisms. All of them possess chlorophyll *a* and at least one other chlorophyll type (*b*, *c*, or *d*) contained in the chloroplast (Table 4-1). This is in marked contrast to the cyanobacteria, which contain only chlorophyll *a* found throughout the cell. Algal cells may be motile by one or more flagella (a trait of the protozoan Mastigophora), exhibit gliding motility as in the diatoms, or are nonmotile. Many of the flagellated types possess a more or less prominent red eyespot.

Algae are important primary producers in fresh and marine waters, soil surfaces, symbioses (e.g., lichens), and a few special environments such as hot springs, saturated salt lakes, etc. They are important oxygen-generating organisms and constitute a significant portion of the plankton in water. As primary producers, they also serve as the start of the food chain for other organisms. A very simple classification used by aquatic biologists (Table 4-2) divides the algae into pigmented flagellates, nonmotile forms, and diatoms.

Protozoa are divided into four groups based on the method of motility. The **Mastigophora** are motile by means of one or more flagella (the algal flagellates; e.g., *Giardia lamblia* causing giardiasis); the **Sarcodina** by means of amoeboid movement (e.g., Entamoeba histolytica causing amebic dysentery); the **Ciliates** by means of shortened modified flagella called cilia; and the **Sporozoa**, which are nonmotile and entirely parasitic (e.g., *Plasmodium* spp causing malaria). Protozoa occupy almost all the habitats in which algae are found as well as many others where light is absent. Protozoa are considered the primary consumers, eating algae, bacteria, and organic particles. Some are symbionts (e.g., in cockroach and termite intestines), a few are pathogens with one group entirely parasitic (Sporozoa), causing many human and animal diseases.

Because protozoa and algae are best observed as live specimens, this presents a few problems: speed of movement and time. For examination of live specimens, especially if flagellates or ciliates are present, it is necessary to slow movement. This is done by adding a small amount of a viscous material to the slide preparation. A ring of **methylcellulose** (10%) is made on the slide and a drop of the sample placed in the center. The methylcellulose diffuses into the sample after the coverslip is in place and slows the motile organisms sufficiently for study.

The second problem is the time required to make observations. A large number of samples may require days to complete. Preservatives may be used in such cases to fix the organisms for later observation. Most features remain intact in these agents. Several are recommended: a water-95% alcohol-formalin (6:3:1) solution is added in equal quantity to the sample; 1 ml of Lugol's iodine (60 g KI and 40 g I_2 per L) is added per 100 ml of sample and stored in the dark.

This exercise is intended to demonstrate a few of the many morphological forms of parasitic and free-living algae and protozoa. As with the other organism groups in this section, specialized courses in phycology and protozoology are usually available.

The prepared slides include four representatives of the many human and animal pathogens among the protozoa. **Malaria** is the most important, considered by some to be the world's most serious infectious disease. There are four types caused by different species of *Plasmodium*. The most serious is *P. falciparum*, which is often fatal and is now showing drug resistance. *P. vivax*, *P. ovale*, and *P. malariae* are less severe. Mixed infections are not uncommon. Animal malaria is caused by other species. Malaria infection begins when infective *Anopheles* mosquitoes bite a human (Figure 4-1). Sporozoites in the saliva of the mosquito are injected into the bloodstream. These migrate to the liver, invading parenchymal cells where they undergo schizogony, forming merozoites which are released into the blood. These penetrate red blood cells undergoing further schizogony, releasing more merozoites and repeating this process again and again. It is the rupture and release of the merozites that produce the symptoms of malaria. The process is not continuous but cyclic, resulting in periods free of symptoms followed by symptoms again. The period between these is in part a function of the parasite species and the feeding habits of the particular *Anopheles* species involved. Some of the merozoites undergo gametogenesis to become gametocytes, which are infectious to the mosquito. In the mosquito stomach, sperm and ova develop, fertilizing each other to form oocysts on the stomach wall. Sporozoites are formed within the oocysts. Upon rupture of the oocyst, the sporozoites migrate to the salivary glands to begin another cycle. The sexual cycle takes place in the mosquito, while the asexual cycle occurs in humans. Diagnosis is made by observing ring forms, mature trophozoites, or gametocytes in red blood cells.

Trypanosoma gambiense causes **trypanosomiasis** or African sleeping sickness. It is transmitted by the bite of the tsetse fly and is often fatal. Several forms occur in cattle and wild animals with some transmissible to humans. American trypanosomiasis (Chagas disease) is caused by a different species and is transmitted by triatomid (kissing) bugs.

Giardiasis is the diarrheal disease caused by *Giardia lamblia*. This disease is associated with drinking untreated water in mountainous areas of the USA or failures in treatment of domestic water supplies. The parasite infects a number of animals other than humans, including beaver and possibly sheep and cattle.

Trichomonas vaginalis is a common cause of a form of **vaginitis** in females and occasionally causes urethritis in males. It is a sexually transmitted disease.

Several free-living algae and protozoa are illustrated in Figures 4-2 to 4-6.

An excellent reference for symptoms, occurrence, treatment, and other aspects of these diseases is *Control of Communicable Diseases in Man*, A.S. Benenson, ed., 16th edition, 1995, American Public Health Association, Washington, DC.

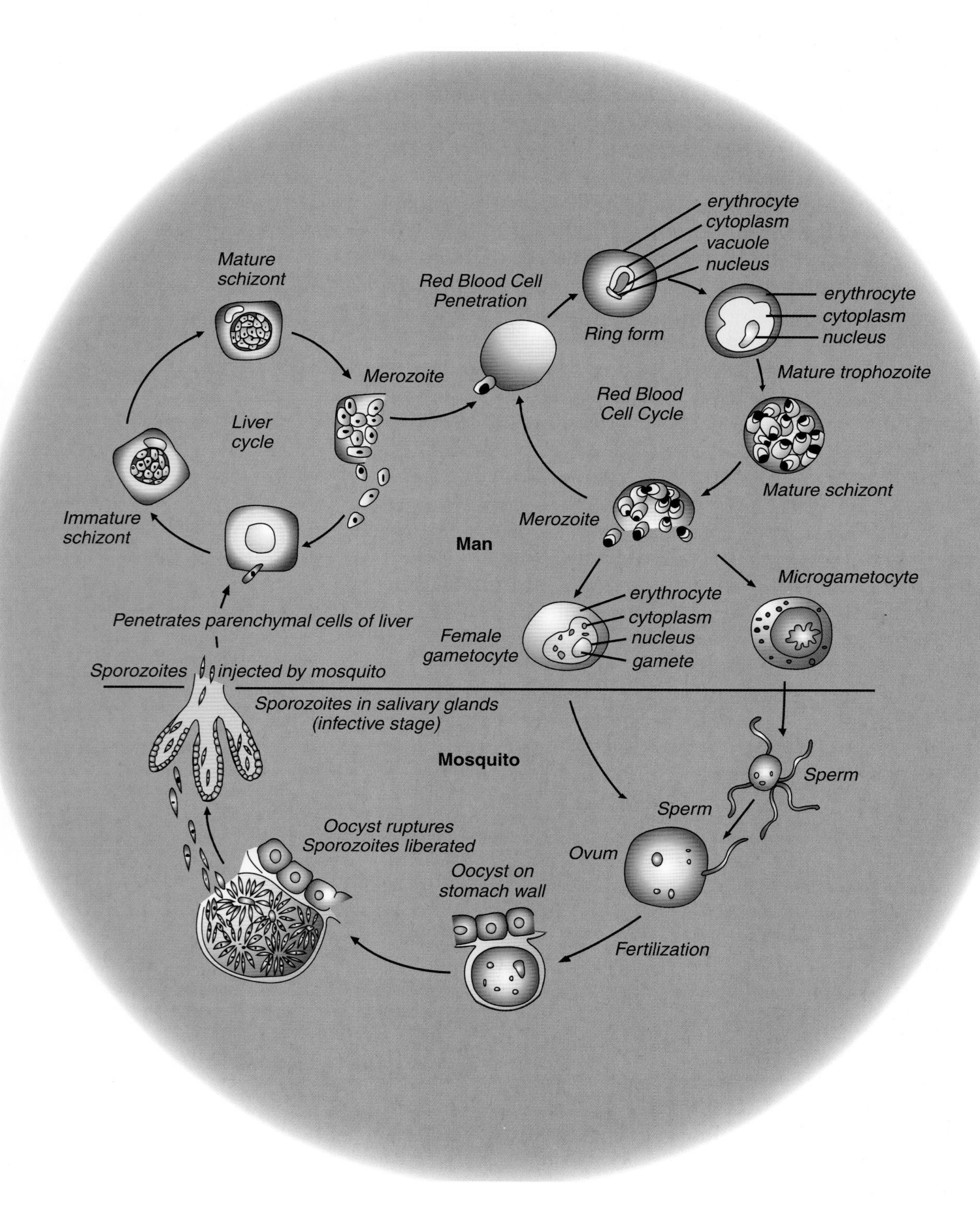

Figure 4-1

Life cycle of the malaria parasite

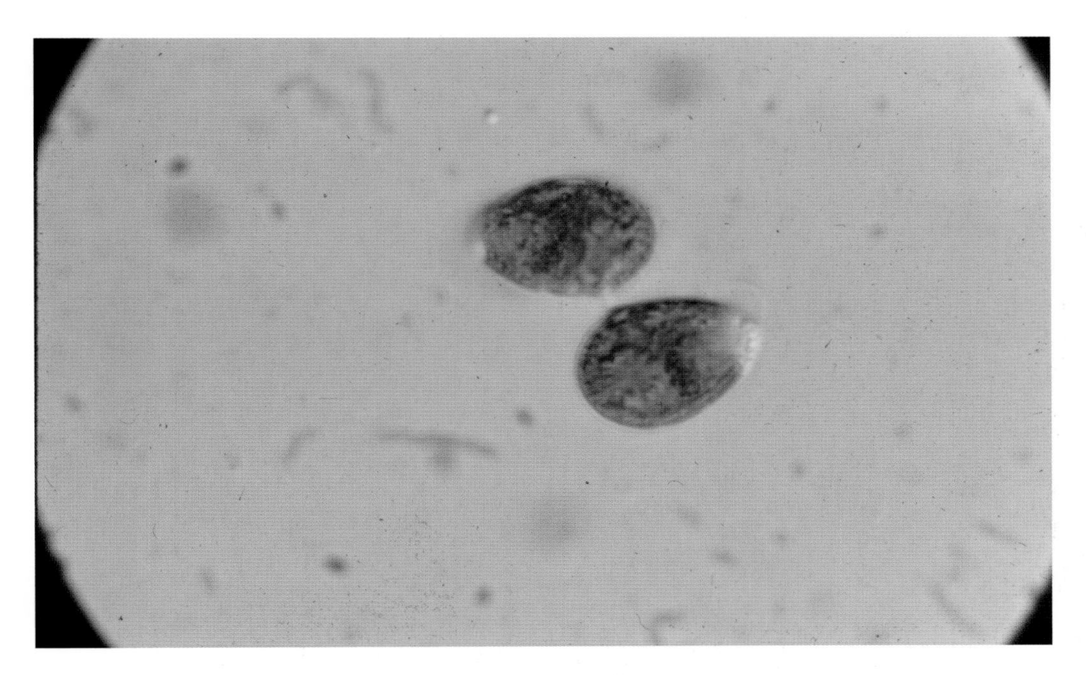

Figure 4-2

Dunalliela salina asexual division. Note division tube, 17 μm long.

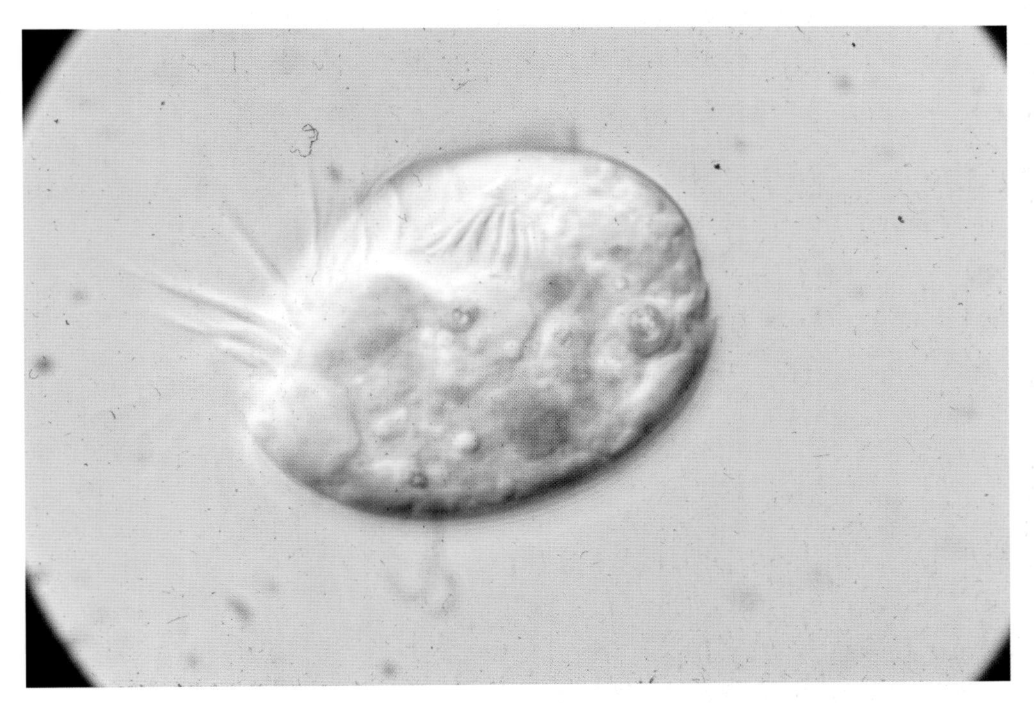

Figure 4-3

Euplotes sp., 43 μm long

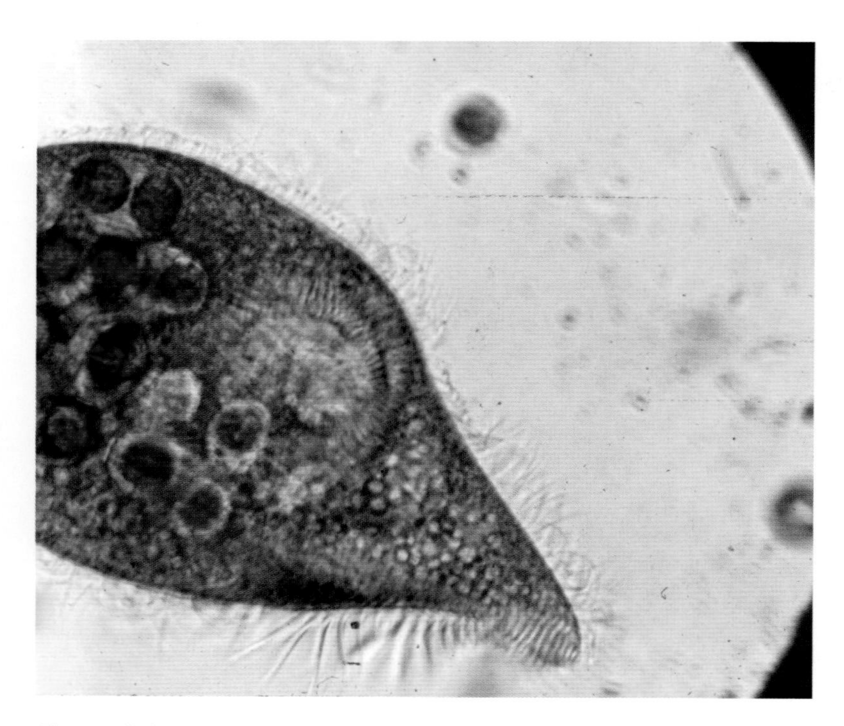

Figure 4-4

Fabrea salina with ingested *Dunalliella salina*, about 100 μm long

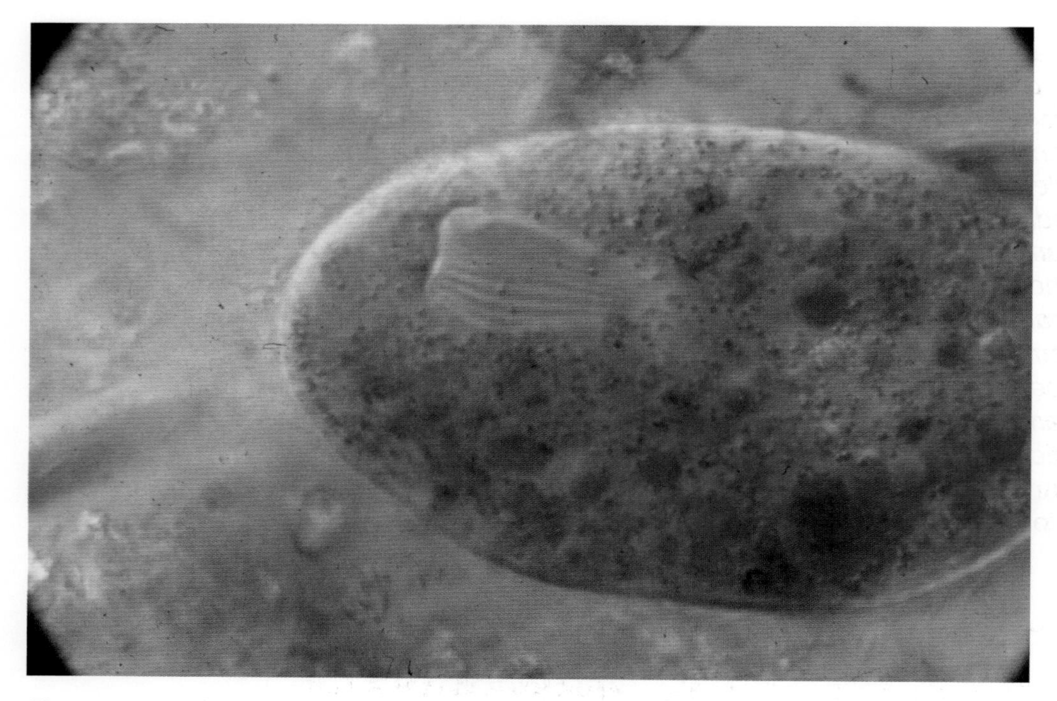

Figure 4-5

Nassula sp., about 100 μm long. Note striated gullet.

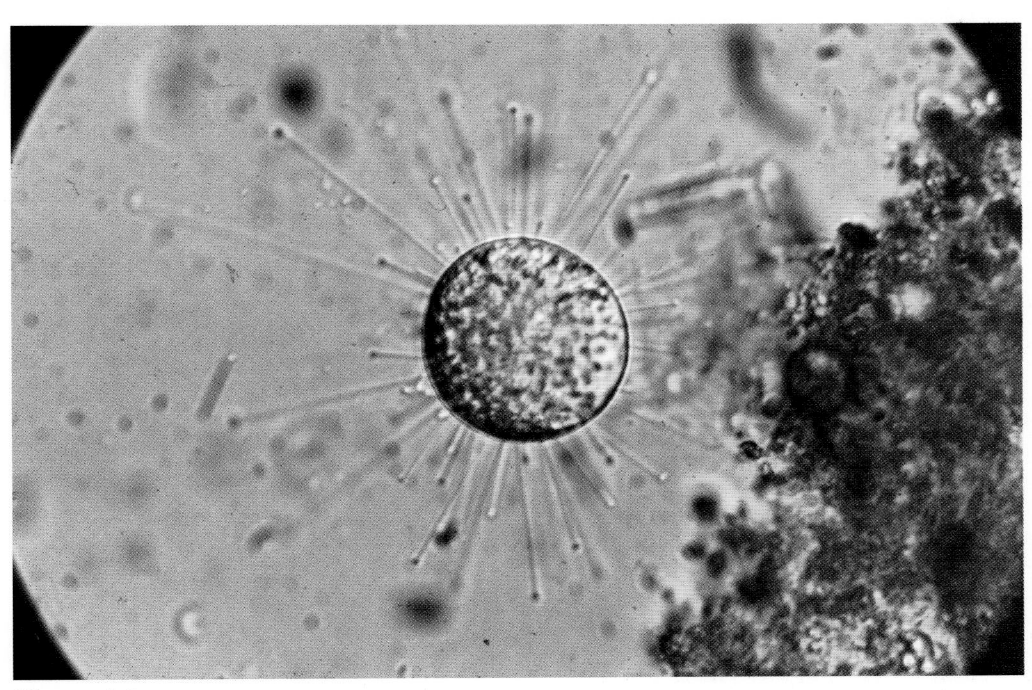

Figure 4-6

Podophyra sp. Note extendable tentacles, 47 μm diameter.

Materials

Part A

1. Prepared slides of a selection of protozoa and algae.
   Synedra, diatom (Chrysophyta)
   Spirogyra, filamentous (Chlorophyta)
   Chlamydomonas (Chlorophyta-Mastigophora)
   Euglena (Euglenophyta-Mastigophora)
   Giarda lamblia, intestinal (Mastigophora)
   G. lamblia, cysts
   Trichomonas vaginalis, vaginitis (Mastigophora)
   Trypanosoma gambiense (Mastigophora)
   Plasmodium falciparum (Sporozoa)
   Amoeba proteus (Sarcodina)
   Paramecium caudatum (Ciliata)
   Vorticella sp. (Ciliata)
2. Malaria life cycle chart (Carolina Bioreview Sheet No. 42-4170) and/or other illustrations.

Part B

3. Live cultures may be available. These will be selected by your instructor.
4. Pond or other sample with live algae and protozoa (seasonal)
5. Slides and coverslips, if live cultures available
6. Methylcellulose, 10% in dropper bottle for live cultures
7. Calibrated oculars, if available

Procedure and Observations

Period 1

A. Prepared slides

1. Place the prepared slide on the microscope stage and focus as described in Exercise 2 using the low-power objective.
2. Depending on the specimen, switch to the high-power objective, focus, and, if necessary, go to the oil immersion lens.
3. Make drawings and labels on the report form in the space provided. Additional paper may be used.
4. It is usually difficult to find the stages of the life cycle on the malaria slides (*Plasmodium* sp.). The Bioreview sheets or other aids supplied by the instructor should help in this.
5. If calibrated oculars are available, specimens may be measured.

B. Live specimens

1. Prepare wet mounts of each organism or the pond sample as described in Exercise 3, Part III.
2. If the organisms are actively motile, place a thin ring of methylcellulose on the slide and add a drop of specimen to the center. Then add a coverslip. In these preparations, oil immersion will usually not be useful.
3. Observe specimens with the most suitable objectives, as needed.
4. Make drawings on the report form and measure with the calibrated ocular, if available.

Table 4-1

Major algal groups

Group Name	Pigment System Chlorophylls	 Others	Cell Wall Composition	Reserve Material	Starch I_2 Test
Brown algae Phaeophyta	a, c	carotenoids	cellulose and align	lamarin and fats	–
Diatoms Chrysophyta	a, c	carotenoids	silica in diatoms	leucosin and oils	–
Dinoflagellates Pyrrophyta	a, c	carotenoids	starch and cellulose	starch and oils	–
Euglenids Euglenophyta	a, b		no wall (pellicle)	paramylum and fats	–
Green algae Chlorophyta	a, b		cellulose	starch	+
Red algae Rhodophyta	a, +/–d	phycobilins	cellulose	starch-like	–

Table 4-2

Comparison of three major groups of aquatic algae

	Pigmented Flagellates	Green Algae	Diatoms
Feature	All Pyrrophyta All Euglenophyta All Chlorophyta and Chrysophyta with flagella	Nonmotile Chlorophyta and Chrysophyta	Bacillarophyceae
Color	Green (brown)	Green	Brown, light green to yellow-green
Starch	Present or absent	Present	Absent
Motility	Flagella	Nonmotile	Gliding in many
Flagellum	Present	Absent	Absent
Cell wall	Thin or absent	Semi-rigid smooth or with spines	Very rigid with regular markings
"Eye" spot	Present	Absent	Absent

5 Helminths

Objectives

The student will be able to:

1. describe the nature of helminths and explain why it is a topic in a microbiology course.
2. define *cestode*, *trematode*, *proglottid*, *scolex*, *ovum*, *monoecious*, *dioecious*, *cysticercus*, *miracidium*, and *cercariae*.
3. diagram and discuss the life cycle of the parasites in schistosomiasis, clonorchiasis, taeniasis, hookworm, pinworm, and trichinosis.

The name **helminth** (helminthology = study of) is a collective term including several metazoan animal phyla of medical and veterinary importance. Members of the phylum **Platyhelminthes** (flatworms) cause the commonly called **fluke** (Trematode) and **tapeworm** (Cestode) diseases, while members of the phylum Nematoda are the causative agents of **roundworm** diseases. These parasites cause disease in most vertebrates including humans. Although rare in the USA, helminth disease is a serious problem in tropical areas of the world. Most cases in the USA are seen in recent immigrants, migrant workers, foreign students, and travelers returning from tropical areas. Helminth disease occurs worldwide in animals other than humans and is frequently a problem in domestic animals.

Helminths are included in microbiology courses, because identification of these agents is done in clinical laboratories by microscopic examination of stools, body fluids, and biopsy. Helminthology is a broad subject and cannot be examined in much detail here. As with many of the microbial taxonomic groups, special courses are available in this subject. The objective here is to introduce the student to a few of the many important helminth agents of human disease.

An excellent reference for symptoms, occurrence, treatment and other aspects of these diseases is *Control of Communicable*

Diseases in Man, A. S. Benenson, ed. 16th edition, 1995, American Public Health Association, Washington, DC.

Trematoda

Flukes are generally referred to as intestinal, liver, blood, or lung flukes because of their primary site of residence as adults. Generally the flukes are monoecious (one animal is both male and female), except the blood flukes which are dioecious (separate male and female animals). Many require an alternate host to complete the life cycle. We will be able to study only a few of the many important species in this group.

Schistosomiasis is the disease caused by blood flukes, members of the genus *Schistosoma.* This is one of the world's most important infectious diseases, perhaps second only to malaria in seriousness and the number of persons infected. The life cycle (Figure 5-1) of the human schistosomes (except for the host, others are nearly identical) begins with the formation of an **egg** which is deposited in the venules of the intestinal wall, bladder, and other sites. These work their way into the feces or the bladder. The eggs have a hook (Figure 5-1A) which often causes damage to the intestinal wall or bladder and urethra and in the latter case results in bloody urine and pussy discharge, a condition illustrated in early-dynasty Egyptian hieroglyphics. Once excreted into water, the eggs hatch into a ciliated intermediate stage called a **miracidium** (Figure 5-1B). The miracidium is infectious only to certain species of aquatic snails in which it undergoes transformation to a **sporocyst** (Figure 5-1C). Daughter sporocysts periodically give rise to vast numbers of free-swimming, fork-tailed **cercariae** (Figure 5-1D). Upon contact with human skin (Figure 5-1E), the cercariae burrow through the skin to the capillaries and then migrate via the bloodstream into the mesenteric vessels and intestinal wall and liver. There they develop into **adult** male (large with a ventral groove) and female (small and narrow) animals (Figure 5-1F). As depicted in Figure 5-1, transmission of the human schistosomes is by contact with water containing human waste and aquatic snails. All water contact—swimming, wading, working, washing clothes, drinking, etc.—are methods of acquiring an infection. Travelers should be wary even of putting their hands in water in areas known to be endemic in this disease. A few animal schistosome cercariae can penetrate the human skin but fail to develop further. A second exposure leads to an immune response resulting in intense itching, the so-called swimmer's itch. This occurs widely in lakes along the Canadian–U.S. border. Diagnosis is made by finding typical eggs in the feces, urine, or biopsy. A serological test is also available.

Clonorchis sinensis, the Chinese liver fluke (Figure 5-2), is an important parasite in the Orient and is often seen in immigrants from that area. This fluke requires two intermediate hosts. Eggs are shed via the bile duct to the feces and then excreted. An egg hatches when ingested by certain aquatic snails producing a sporocyte which releases a large number of cercariae. These burrow into certain freshwater fish and encyst in the muscles. When raw or undercooked fish is ingested by humans or other fish-eating mammals, they excyst and migrate to the liver via the bile duct where they develop into adults. Diagnosis is made by finding eggs in feces.

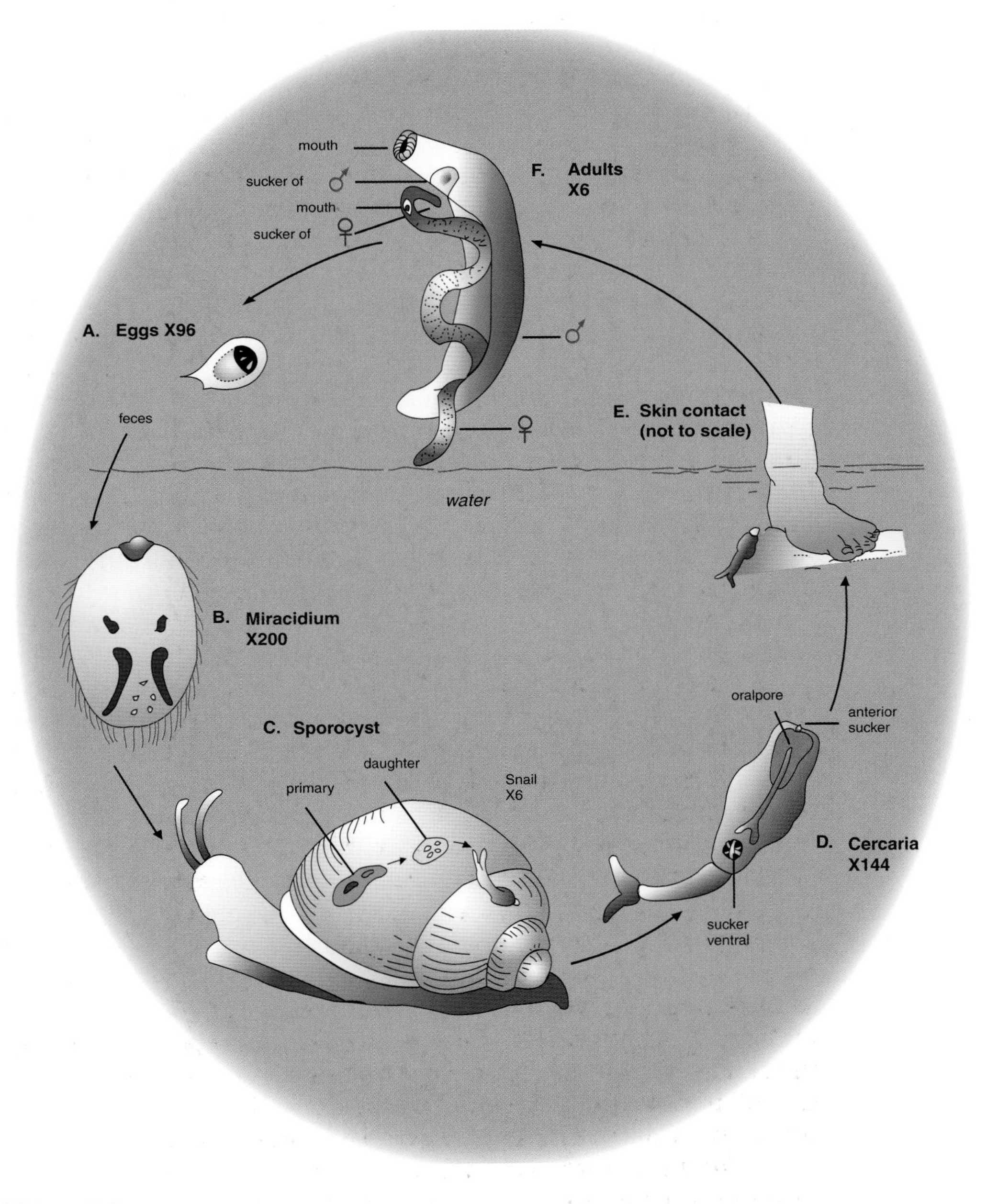

Figure 5-1

Generalized life cycle of the schistosomes

Cestoda

Tapeworms (cestodes) consist of a scolex or head with or without hooks to attach to the intestinal wall. The head produces a long chain of proglottids, each being monoecious, which break off the terminal end as they reach maturity and are shed in the feces (Figure 5-3). An intermediate host is usually involved.

Taenia solium, the pork tapeworm, has 22–32 hooks on the **scolex** (Figure 5-3B) with a distinctive **proglottid** (Figure 5-3A). The pig is infected when eggs are ingested with food contaminated with human feces. The eggs hatch into a stage which migrates through the intestinal wall and via the blood to the liver where further development takes place. The cysticerci leave the liver and enter the mesenteries and muscles where they encyst. Humans are infected by eating undercooked or raw meat. The cysts are released in the intestinal tract during digestion where they develop into adults (Figure 5-3C). In contrast to *T. saginata*, the beef tapeworm, humans can be infected by *T. solium* eggs. When the cysticerci encyst in muscles, they do little harm; however, when they encyst in nervous tissue or sense organs, they can cause considerable damage, even causing death if in the brain. Diagnosis is made by finding characteristic proglottids or eggs (rarely in the case of *T. solium*) in feces. Finding the scolex after treatment is definitive.

Diphyllobothrium latum, the fish tapeworm, occurs widely in North America. Hosts include bears, cats, dogs, humans, and other fish-eating carnivores. Human infections have produced tapeworms up to 60 feet in length. Eggs produced from mature proglottids in the mammal host hatch when shed into water producing a ciliated stage infectious to certain copepods in which they develop. When the copepod is eaten by certain freshwater fish (pike, walleye, salmon, perch, turbot, and others), this stage burrows through the intestine to reach the body wall or viscera where further development takes place. When raw or undercooked fish is eaten by the final host, this stage develops into the adult form in the intestine. Eggs are formed in the proglottids, released, and shed with the feces, thus completing the life cycle. Diagnosis is made by finding proglottids and eggs in the feces.

Nematoda

Roundworms of the phylum Nematoda are responsible for a number of very important human and animal diseases. Among them are pinworm, hookworm, ascariasis, river blindness (onchocerciasis), trichinosis, whipworm, heartworm of dogs, filariasis, anisakiasis, and loa loa, to name a few. Intermediate hosts and vectors are sometimes involved.

Pinworm (*Enterobius vermicularis*, Figure 5-4) is perhaps the most common helminth infection in the USA, principally in

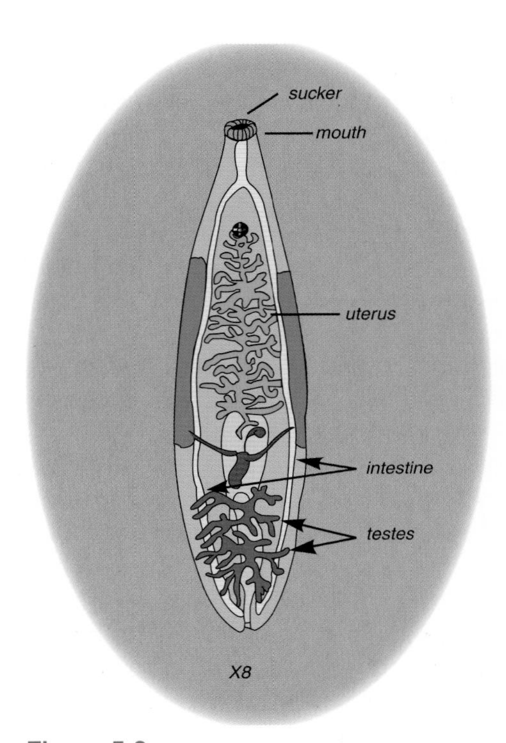

Figure 5-2

Clonorchis sinensis

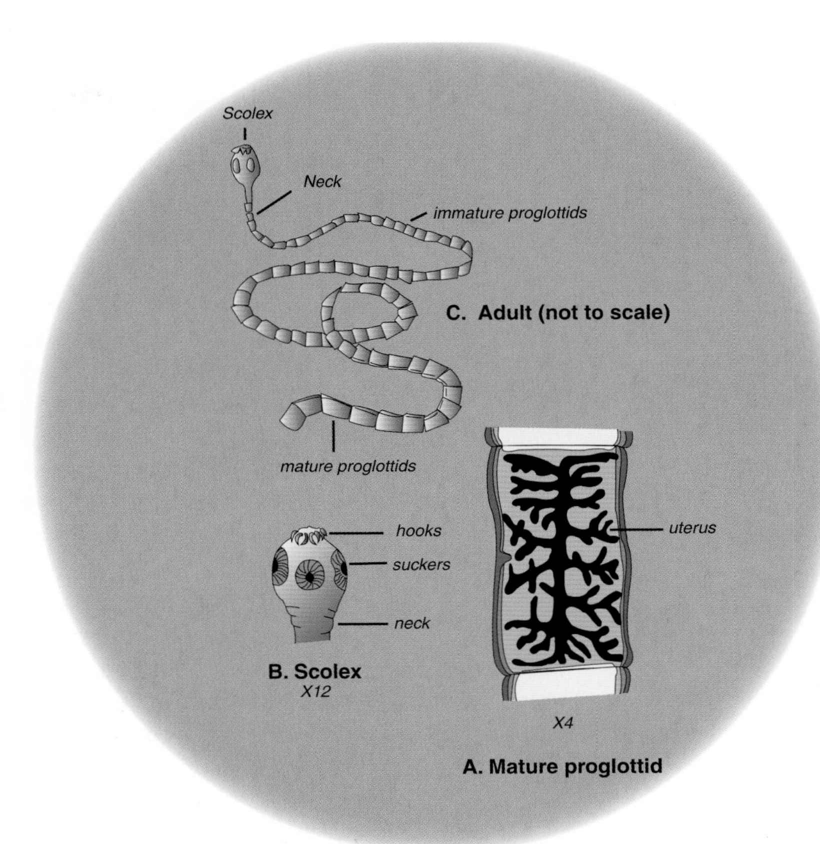

Figure 5-3

Taenia solium. Adult scolex is about 1 mm diameter and length may reach 15–16 meters

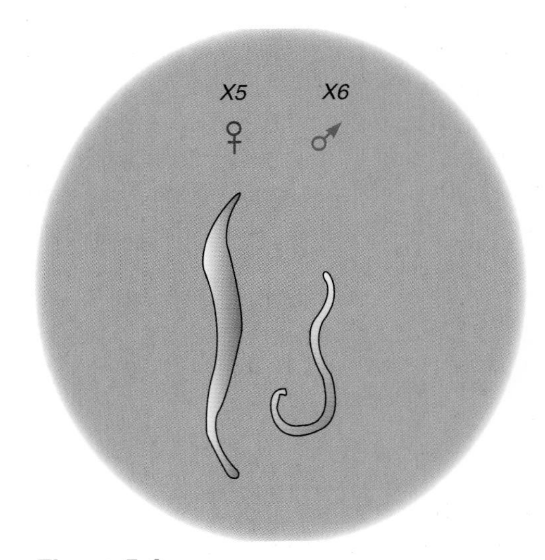

Figure 5-4

The pinworm, adults

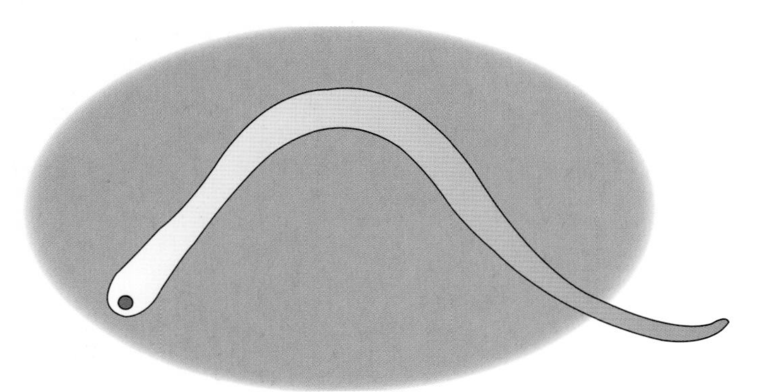

Figure 5-5

The hookworm, adults (up to 8 mm long)

school-age children (kindergarten to 3rd grade) and secondarily in their immediate family members. The adult worm lives in the intestine, migrating to the anus during the night to lay eggs. The anal area is often irritated, causing the child to scratch the area continuously, interrupting sleep and causing general irritability. The eggs are passed to the clothing, bedding, and generally to house dust. Ingestion of eggs from dust, fingers, etc., is relatively easy, often infecting all household members. The eggs hatch into larvae in the intestine, where they develop into adults. The finding of eggs or adults in the anal area is diagnostic.

Hookworm disease is common in Africa (*Ancylostoma duodenale*, Figure 5-5) and parts of the Western Hemisphere including the southern states in the USA (*Necator americanus*). The disease is frequently characterized by chronic anemia due to loss of blood, especially in children. Eggs are deposited on moist ground and hatch, passing through several larval stages in the soil. This development requires warm, moist, and humid conditions and the disease is limited to tropical areas of the world as a result. Humans are infected when the larvae penetrate the skin, usually through bare feet, and cause dermatitis (ground itch). The larvae pass via the blood and lymphatic system to the lungs, migrating up the larynx to the trachea whereupon they are swallowed. Reaching the intestine, they attach to the wall and develop to maturity, producing eggs in 6–7 weeks. Diagnosis is by observing eggs in feces.

Trichinosis (*Trichinella spiralis*) is a roundworm parasite of carnivorous and omnivorous animals. The larvae encyst in muscle tissue (Figure 5-6), and it is transmitted by eating raw or undercooked meat. In humans, pork and pork products are the most common vehicles. In the USA, bear meat is often the source. Larvae develop into adults in the small intestine mucosa. Fertilized females then produce larvae which penetrate the lymphatics and are disseminated throughout the body. Larvae migrate into muscle and encyst. Finding larvae in muscle biopsy is the usual diagnostic procedure. A serological test is also available.

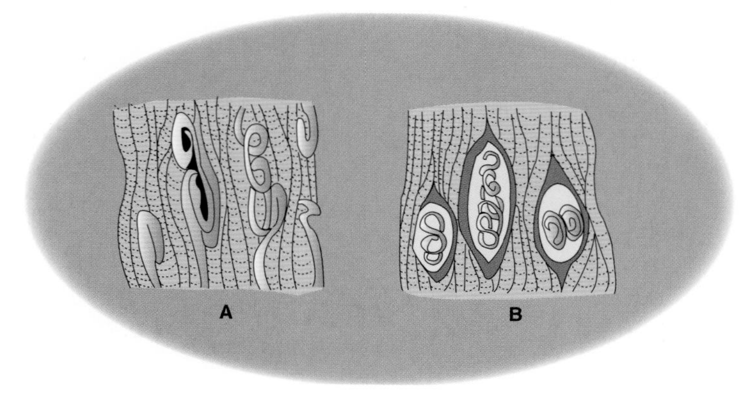

Figure 5-6

(A) *Trichinella spiralis* larvae migrating into human tissue; (B) encysted larvae in porcine muscle (X60)

Anisakiasis is caused by several genera of the family Anisakidae including *Anisakis*. It is transmitted by eating raw or inadequately treated (salted, frozen, smoked) saltwater fish or squid (sushi, sashimi, ceviche) and herring. The larvae burrow into the intestinal wall causing nausea, vomiting, diarrhea, and pain. The larvae are about 2 cm long and often migrate into the throat, where they can be removed with the fingers. Diagnosis is by recognition of the larvae removed from the throat or gastroscopic examination.

Materials

1. Prepared slides of a selection of helminths:

Organism	***Ova***
Schistosoma mansoni, female	*Schistosoma mansoni*
Schistosoma mansoni, male	
Schistosoma mansoni, cercariae	
Schistosoma mansoni, miracidium	
Clonorchis sinensis	*Clonorchis sinensis*
Taenia solium, scolex	*Taenia solium*
Taenia saginata, proglottid	
Taenia solium, cysticercus	
Diphyllobothrium latum, scolex	*Diphyllobothrium latum*
Enterobius vermicularis	*Enterobius vermicularis*
Necator americanus	*Necator americanus*
Trichinella spiralis, encysted larvae	
Trichinella spiralis, migratory larvae	

2. Preserved specimens may also be made available
3. Microscopes
4. Stereo dissecting microscopes

Procedure and Observations

Period 1

1. Examine the prepared slides using either the low-power objective of your microscope or the dissecting microscope. Your instructor will indicate which is most appropriate for each specimen.
2. Compare what you see with the drawings in this exercise. Eggs are shown in Figure 5-7.
3. Make a drawing of what you see. Label structures as far as you can determine them. It is important that you be able to recognize the parasite later.
4. Write a short description of each disease and its epidemiology in the space provided on the report form.

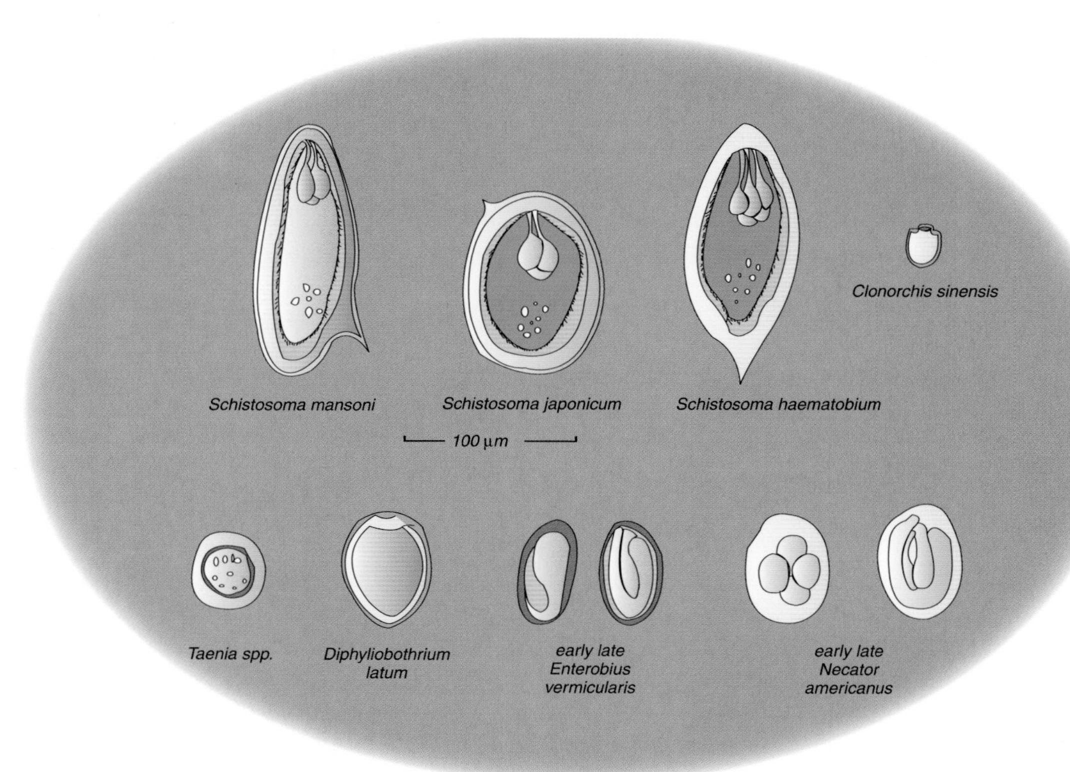

Figure 5-7

Helminth eggs, all to same scale

UNIT II

Aseptic Technique, Cultivation, Isolation, and Counting of Microbes

This unit deals with the various methods of cultivation, handling, and enumeration of microbes. A fundamental consideration of this section and of all areas of science involving microbes is aseptic technique. The word aseptic is derived from the Greek *septikos*, to make putrid. The word septic is defined as a practice that promotes putrefaction. The prefix a- changes the meaning to "without"; thus **aseptic** means a practice that prevents putrefaction. In microbiology and medicine, it refers to the handling of microbes, wounds, media, and related material in a manner to prevent the introduction or growth of unwanted organisms. Aseptic technique is conceptually easy to understand but sometimes difficult to practice. It requires an understanding of the sources of microbes in the environment and rigorous attention to what one is doing until its practice becomes second nature. The student will be exposed to aseptic techniques in each of the exercises of this unit and attention will be called frequently to steps with potential problems.

In order to treat disease, learn about the spread of pathogens (epidemiology), or to genetically engineer organisms, one must first isolate the organism of interest as a pure culture. The German bacteriologist, Robert Koch, was the first to put pure culture rules into practice in demonstrating that a particular organism is the cause of a specific disease. Koch (1) isolated an organism from a disease process as a pure culture; (2) it was always associated with the particular disease; (3) the organism was then injected into a test animal resulting in the same disease; and (4) the same organism was reisolated as a pure culture and demonstrated to be the same as the first organism isolated. These rules have come to be known as **Koch's Postulates**. In this unit the student is introduced to the preparation of media for growing organisms, the isolation of organisms, the recognition of a pure culture, and the observation of growth patterns while practicing aseptic technique. Finally, a method will be introduced for counting and measuring organisms in a culture.

In some teaching situations, Exercise 6 may not be done by the students. The material in the introduction to that exercise is still a necessary part of the knowledge of every microbiology student and should be assigned reading in any case.

6
Media Preparation and Sterilization

Objectives

The student will be able to:

1. prepare broth and agar media, dispense into tubes, and sterilize by autoclaving.
2. describe the consequence of failing to sterilize the media.
3. discuss the origin of the organisms resulting in growth in the unsterilized media.
4. define the following terms: *nutrients*, *macro-* and *micronutrients*, *culture medium* and its variations (*synthetic*, *complex*, *differential*, *selective*, *minimal*, *enriched)*, *agar*, *sterilize*, *autoclaving*, *pure culture*, and *aseptic technique*.
5. distinguish between liquid and solid media.

(**NOTE:** Even if this assignment is not done in the laboratory, the student will still be responsible for the material presented here.)

Nutrients are chemicals that microorganisms use to supply energy and to build cell material ultimately leading to division. **Macronutrients** are those required by the cell in fairly large amounts, since they are sources of energy or are used directly in cell material. These include carbon-, nitrogen-, and sulfur-containing compounds, phosphorus, magnesium, iron, sodium, and potassium. **Micronutrients** are those used in extremely small but essential amounts and are often cofactors in enzymes. These include cobalt, manganese, molybdenum, zinc, and, in some cases, vitamins. Many of the micronutrients may be found as contaminants of macronutrients, but with some organisms they must be supplied specifically. Every organism has a requirement for specific nutrients; some may be very fastidious, requiring a considerable variety, while others may require only a few chemicals.

A **culture medium** (pl., **media**) for growing an organism may be **synthetic**, where the chemical structure of each ingredient is known; **complex**, where the chemical structure of one or more of the ingredients is unknown; **differential**, where the medium contains a reagent giving different reactions with different organisms (e.g., an acid-base indicator), Figure 6-1; or **selective**, where the medium contains an agent allowing one organism to grow while inhibiting others (Figure 6-2). Media may also be described as **minimal**, containing just the bare minimum for growth; or, **enriched**, with large amounts of one or more ingredients to encourage maximum cell yield. Despite all the available ingredients, many organisms from nature will not grow in a particular medium, or, indeed, any at all. The medium may be too rich, not rich enough, or some critical but unknown ingredient is missing. In other words, there is no single medium that will grow all the organisms in nature. Fortunately, a few rich media will grow many of the microorganisms of interest.

To grow a given microbe, the culture medium must contain those nutrients essential for the growth of that organism and provide suitable surroundings for growth—proper pH, osmotic pressure, oxygen, temperature, etc. Many different substances will serve satisfactorily in a culture medium. Essentially, all culture media are one of two forms: (1) **liquid** (broth) or (2) **solid** (agar and other gelling agents, e.g., silica gel, gelatin, etc.). The formulas for culture media are highly varied. Those used in the exercises in this manual will be found in Appendix 1.

By far the most common solidifying agent in use is **agar**, a colloidal gelling agent. It is a polysaccharide obtained most commonly from a marine red alga and available in a dried form of many degrees of purity. Chemically it is a galactan sulfuric acid ester, which is soluble (melts) in boiling water but not in cold water. Only a few specialized marine bacteria can digest agar (i.e., use it as a nutrient), so it is used for its gelling property rather than as a nutrient and can be used in synthetic media for this reason. It is usually added to a broth medium at a concentration of 1.5% (w/v). A concentration of 1.8–2.5% gives a harder, less easily gouged medium, but less water is available for growth and some organisms will not grow on it. Concentrations of 1% or less are often used in specialized media (e.g., motility media).

Although different agars may vary considerably in their physical properties, the usual melting point is 95°–100°C. Thus, solid agar media are liquefied for use by boiling. On cooling, a medium containing agar solidifies at about 42°C. If it is to be inoculated before hardening, it is usually cooled to 45°–47°C, a temperature that is not harmful to many microorganisms for a short time. Once solidified, agar may be incubated over the entire range a microbiologist is likely to use (up to 70°C, perhaps) without melting.

A freshly prepared medium contains numerous microorgan-

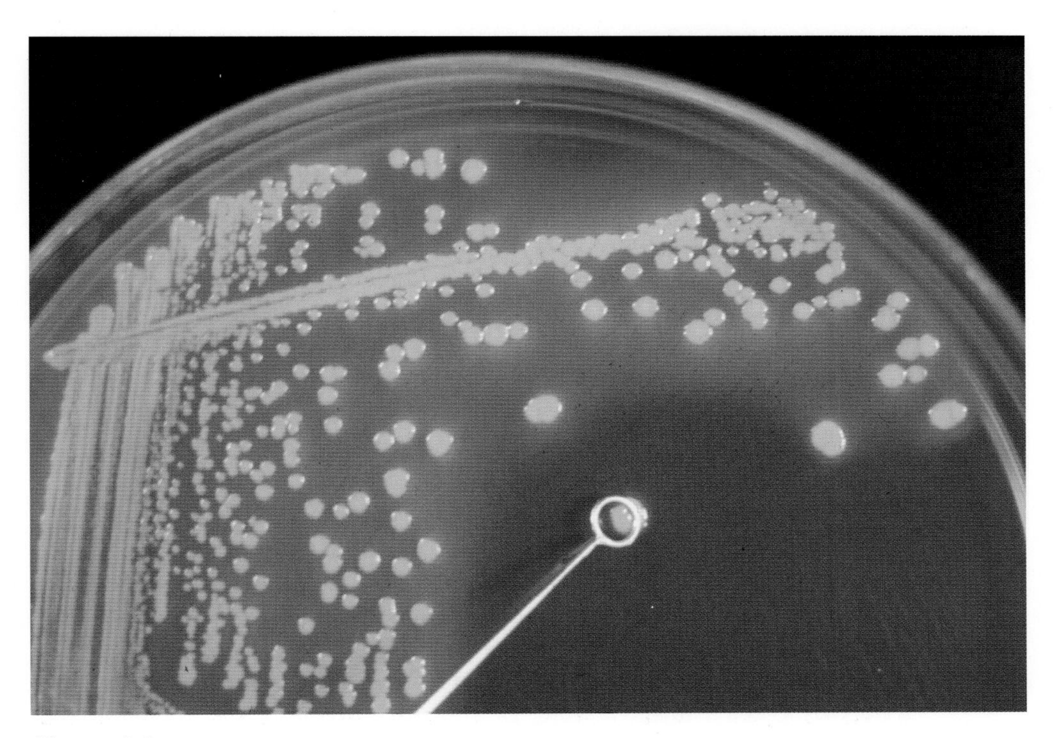

Figure 6-1

A selective differential medium: Endo agar with *Escherichia coli* colonies

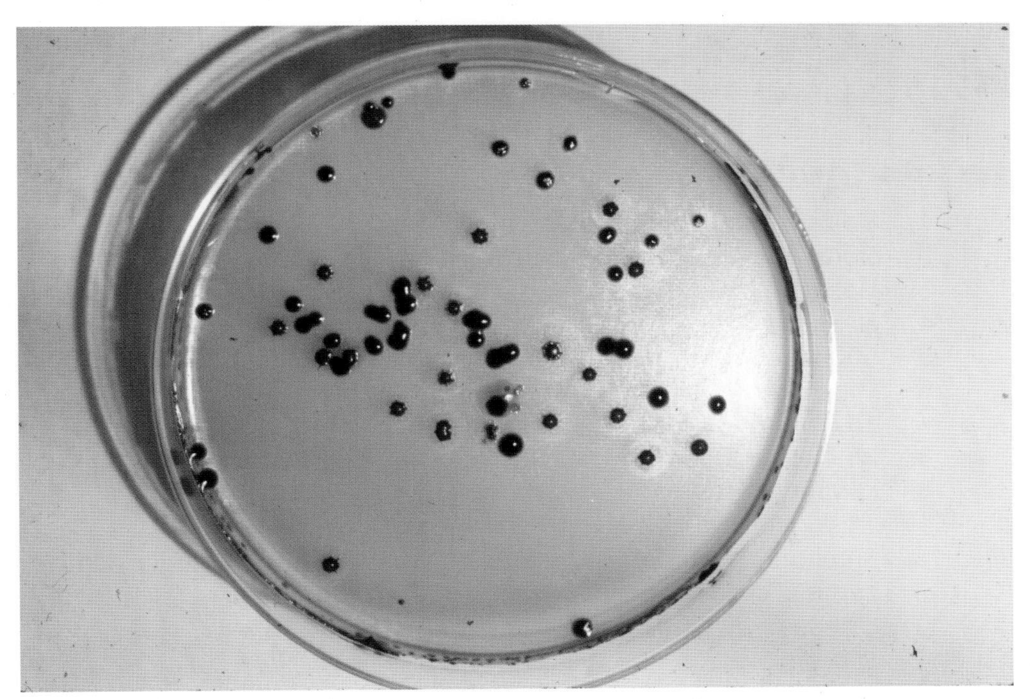

Figure 6-2

A selective medium: Baird Parker agar with *Staphylococcus aureus* colonies

isms found in the ingredients, the water used to prepare it, and from the utensil surfaces and glassware; therefore, it must be **sterilized** (i.e., heated to a point where all organisms present are destroyed). Otherwise, a mixture of unwanted organisms would result. Prior to sterilization, the container is usually plugged with cotton or loosely capped with aluminum foil or other closure. This prevents the entry of new contaminants while permitting free interchange of air or gases.

Media containing moisture, whether broth or agar, are most often sterilized by the use of steam under pressure called **autoclaving** or **pressure cooking**. The temperature of free-flowing steam (and thus boiling water) is 100°C at sea level. However, this temperature is not high enough to sterilize, since only vegetative forms of microorganisms are destroyed and not the endospores of sporeforming bacteria. Applying free-flowing steam at one atmosphere of pressure above normal air pressure increases the boiling temperature of water to 121°C. Maintaining the temperature at 121°C for 15–20 minutes will sterilize the media in most containers of less than one liter volume. Larger volumes require a longer time.

Since most of your laboratory study will be made with **pure cultures** (i.e., a single species of microorganism), a sterilized culture medium must be maintained and manipulated in a sterile condition, free of living forms. You must also be able to inoculate this sterile medium with a pure culture without outside contamination. This is commonly referred to as **aseptic technique**. In this exercise you will have the opportunity to prepare a medium in several forms, sterilize it, and use the product in the succeeding exercises.

Materials (per pair)

1. Bottle of nutrient broth (dehydrated)
2. Bottle of bulk agar (plain, dehydrated)
3. Balance, triple beam
4. Spatula
5. 500 ml beaker, washed but not sterilized
6. Glass rod for stirring
7. Test tube rack with 22 mm diameter holes
8. 14 culture tubes, 16 mm diameter, washed but not sterile and not capped
9. 14 caps or closures for 16 mm tubes
10. 6 culture tubes, 20 mm diameter, washed but not sterile and not capped
11. 6 caps or closures for 20 mm tubes
12. 1 10 ml pipet
13. 1 pipet aid for 10 ml pipets
14. Weighing papers or cups
15. Graduated cylinder with 250 or 500 ml capacity

Procedure

Period 1

A. Broth preparation

1. Place a weighing paper on the balance pan and weigh it by moving the weights until the arm points to the zero mark. Read the weight to the nearest 0.1 g and record it on a piece of paper.
2. Add 20.0 to the weight recorded on the paper and change the weight positions on the balance beams until the total weight corresponds to your calculated weight.
3. With a distilled water-rinsed and dried spatula, add small amounts of the powdered nutrient broth to the weighing paper until the balance arm is again at the zero position. If you add too much, remove a little until it does balance **BUT DO NOT RETURN ANY POWDER TO THE BOTTLE. DISCARD IT INSTEAD**. This is to avoid any possible chemical contamination of the supply bottle. ***Replace*** the cover on the bottle of medium ***promptly*** since the powder is ***very*** hygroscopic. Dump the powder into a 500 ml beaker.
4. In a graduated cylinder, measure 250 ml of distilled or demineralized water and add it to the beaker with the broth powder.
5. With a stirring rod, mix the beaker contents until all of the powder has gone into solution and a clear yellow-brown liquid is present. There should be no cloudiness. The powder tends to form clumps or lumps and solution may take some time. Ordinarily, the pH would be checked and adjusted to about 7.0–7.2 before autoclaving (drops during autoclaving to about 6.8). This won't be done here. Nutrient broth is about pH 6.9 before and 6.8 after autoclaving.
6. Place 8 of the 16 mm diameter culture tubes in a rack. With a 10 ml pipet and aid, pipet 10 ml of the nutrient broth into each of the tubes. Set the pipet aside for later use. The remaining beaker contents will be used to make the agar.
7. Cap the filled tubes with the appropriate sized caps.

B. Agar preparation

1. Weigh out a paper as done in step 1 above.
2. This time add 2.6 grams to the recorded paper weight and change the balance beam weight to correspond.
3. With a distilled water-rinsed and dried spatula, weigh out the agar onto the paper as done for the nutrient broth. This amount of agar will be 1.5% in the remaining volume of broth. As before, **DO NOT** return any excess agar to the supply bottle; discard it.
4. Place the stirring rod in the beaker and place the beaker on a tripod so that it can be heated by a Bunsen burner flame. Note that the broth is now cloudy with granules of agar that settle out quickly. This may be done in a microwave oven on high setting, but the broth must be stirred every minute or so.
5. Begin heating and stir constantly to keep the agar suspended. If it settles for even a short time, it will char and ruin the medium.

6. Keep a constant watch for boiling, stirring all the time. Because agar and broth media have a very low surface tension, boiling occurs ***suddenly*** and is accompanied by rapid foaming, often running over the beaker's edge. This can be an even greater problem in flasks with narrow necks or when done in a microwave oven. When it reaches boiling, remove the flame, but keep the contents boiling gently while stirring until the granular appearance of the medium disappears.
7. Using the 10 ml pipet you previously set aside, pipet 7 ml into each of 6 of the 16 mm diameter tubes and 20 ml into each of the 20 mm diameter tubes. Cap the tubes with the appropriate cap. Separate these tubes from the broth with a penciled note on a piece of paper because you will need to do something with them after sterilization.

C. Sterilizing

1. Remove 2 broth tubes; write your name on each. Place one at room temperature (20°–25°C) and one in the refrigerator (about 2°C) until the next laboratory period. Also place the beaker with the remaining agar at room temperature until the next period.
2. Take the rack with the remaining tubes to the autoclave and sterilize at 121°C for 15 minutes. Your instructor may have to consolidate the various media to conserve space in the autoclave.
3. After autoclaving is complete (about an hour), the 16 mm diameter agar tubes must be removed from the rack (Beware—hot) and slanted. This is done by laying the tubes on a bench with the capped end elevated slightly using stacked paper sheets, a Bunsen burner hose, a wood pencil or other device, to give a liquid level extending from just around the curve of the tube bottom to about 3/4 of the distance to the cap. The "butt" (the portion at the bottom of the tube) should be about 0.5-1 cm up the side (see Figure 22-1 for example). Leave the tubes in this position until the agar solidifies.

Observations

Period 2

1. Take the tube from the refrigerator and the tube held at room temperature and, using one of the sterilized broth tubes as a control, hold all three up to the light. Observe the appearance of the three tubes, noting turbidity (i.e., cloudiness) in the room temperature tube. The refrigerated tube should be as clear as the sterilized tube. Why?
2. Observe the agar in the beaker left at room temperature, noting any colonies on the surface or embedded in the agar itself.
3. Note how the appearance of the agar slant medium has changed since it solidified.
4. Enter your observations on the report form.

7 Broth and Agar Slant Culture

Objectives

The student will be able to:

1. transfer a culture to broth and an agar slant using proper aseptic technique without gouging the agar slant.
2. recognize and describe the various growth patterns in broth culture and on agar slants.
3. describe other growth characteristics such as relative amount, chromogenesis, and consistency.
4. meet Objectives 2–5 of Exercise 6 whether or not that exercise was performed.

(**NOTE:** The student should read and have a thorough command of the introductory material to Exercise 6 before undertaking this or subsequent exercises whether or not Exercise 6 was actually done in the laboratory.)

To grow a microbial culture in or on a sterilized medium, the **inoculum** (cells) is transferred (**inoculated**) to the medium, using special precautions to maintain the purity of the culture being transferred. **Aseptic technique** is the term applied to these precautions.

When a wire needle or loop inserted into a holder is used to transfer microorganisms in the inoculation procedure, it should be heated to redness by flaming immediately before and after making a transfer in order to kill all organisms present. It is important that the lower part of the holder be included, because it also may enter the container of medium and may thus serve as a source of microbes if any are left alive. Hold the loop down the side of the Bunsen burner flame as in Figure 7-1A so that only the upper part of the wire and the holder is heated. Draw the loop

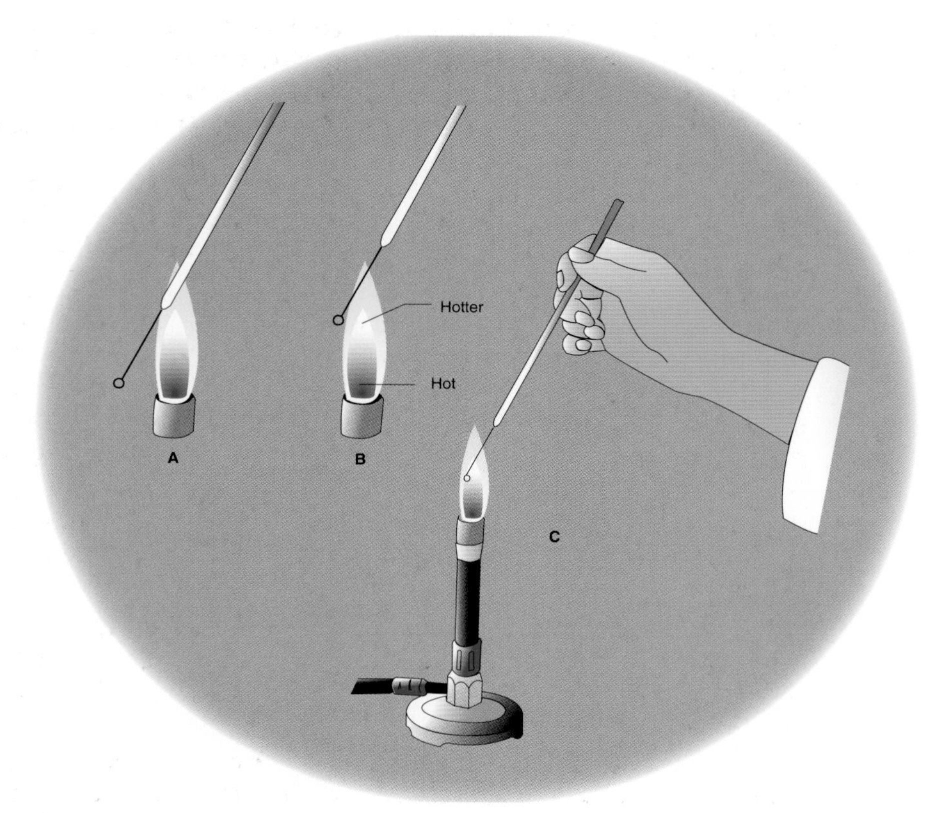

Figure 7-1

Sterilizing the wire inoculating loop in the Bunsen burner flame. Start with the loop outside the flame cone (A), and slowly draw the loop into the hot part of the flame (B), allowing evaporation of liquid before incineration (C). This prevents splattering of culture and cells.

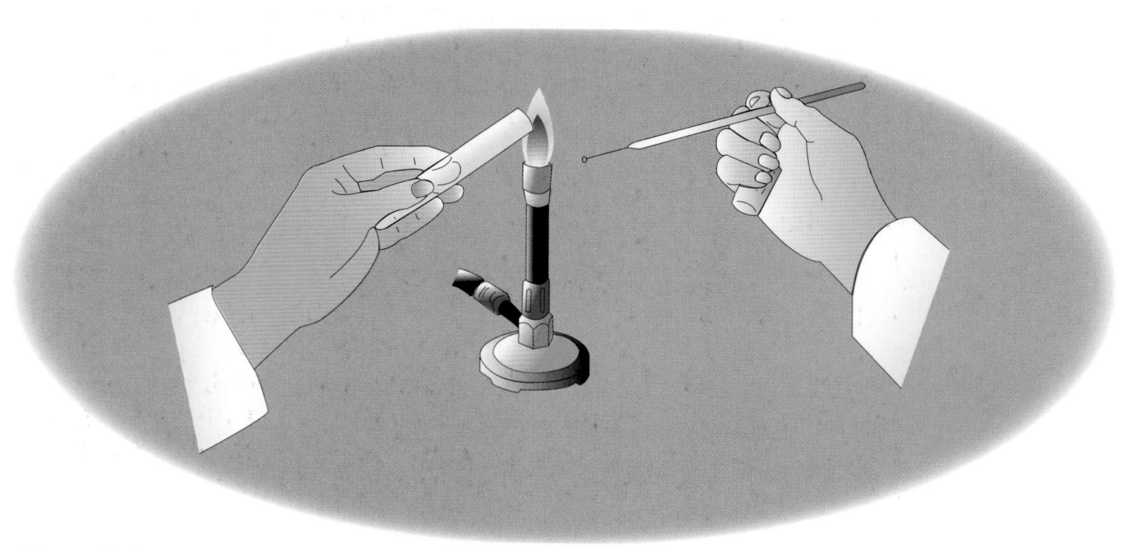

Figure 7-2

Flaming an open culture tube prior to inoculation. The tube is held almost horizontally and the cap held by the little finger of the right hand which holds the inoculating loop.

slowly into the flame as in Figure 7-1B until it is red (Figure 7-1C). This practice is especially important after transferring an inoculum, since it allows the moisture on the loop to evaporate before being incinerated, thus reducing the hazard of splattering due to instantaneous heating.

During the transfer of cells, hold the culture tube in the left hand (if you are right-handed) as nearly horizontal as feasible and grasp the cap between the fingers of the right hand (Figure 7-2) and the palm. **CAUTION:** Never lay a cap down! The mouth of the tubes from which the culture is taken and into which it is transferred must be passed through the burner flame immediately *after removing the cap and before* replacing it. This is done in order to (1) kill any organisms on the lip and (2) heat the glass sufficiently to create an updraft of air preventing airborne organisms from falling into the tube. Don't overheat the glass and don't leave the tube open any longer than necessary. Insert the sterile inoculating wire, touch the culture, and withdraw the wire. Flame the lip of the tube and reseat the cap.

Pick up the tube to be inoculated, remove the cap, flame the lip, and insert the loop or needle with the culture on it. Move the wire back and forth once or twice in broth or up and down on the slant (without gouging the agar), then remove the loop or needle, and flame it as described above. Flame the lip of the inoculated tube and replace the cap. With a little practice, several tubes can be held in one hand and done at the same time. ***Aseptic technique failures often occur at this point***; such as, when attempting to remove the caps the sterile wire comes in contact with your sleeve, the bench top, or other source of contamination; the cap is left off of the tube or tubes too long; the cap is set down on the bench; or too much time is taken with the transfer itself. Since most of a microbiologist's work is done with pure cultures, you should appreciate the importance of the aseptic techniques involved in inoculation and master them early in the course.

Following inoculation, a culture is **incubated** in an environment providing suitable growth conditions. "Growth" here means the development of a population of cells from one or a few cells. The mass of daughter cells becomes visible to the naked eye either as **turbidity** (cloudiness) in liquid broth or as an isolated population (**colony**) on solid media. The visible appearance of growth is sometimes an aid in differentiating microbial species.

This exercise is divided into two parts using broth and agar media, which will provide you with experience in transferring cultures and the accompanying aseptic technique as well as the terms used in observing growth patterns.

Materials (per pair)

1. 18- to 24-hour broth cultures of *Escherichia coli*, *Micrococcus luteus*, *Bacillus subtilis*, *Enterococcus faecalis*, and *Serratia marcescens*
2. 6 tubes of nutrient broth or tryptic soy broth, sterile (from Exercise 6, if done)
3. 6 tubes of nutrient agar or tryptic soy agar slants, sterile (from Exercise 6, if done)

A. Broth Media

The use of broth media is a convenient way to handle bacteria in **stock culture** (i.e., keeping a culture as a source for future subcultures), and organisms often grow in a characteristic manner in the broth medium.

Cells may grow dispersed, showing interruption of the light path or cloudiness called **turbidity**. This may be slight to moderate to heavy. Cells may settle to the bottom to form a **sediment** or **button**. Cells may be attracted to each other to form clumps that settle to the bottom. These clumps may be adherent or slimy when dislodged from the bottom. Organisms that form an extensive capsule may be very viscous or slimy, sometimes drawing out in a long string when touched by an inoculating loop. A small button often may be the only sign of growth. Examine a tube first by looking up from the bottom. Then gently tap the tube near the bottom. A slight sediment will swirl upward. A button is not always growth but may be simply inoculum cells that have settled out. If there is no increase with time, this is probably the case. Many organisms grow in a film across the surface called a **pellicle**, which may be heavy or light and membranous. Some organisms form a **ring** of growth around the glass-broth-air interface, seen by tilting the tube slightly. Handle tubes carefully when first observing them because heavy pellicles often fall to the bottom when disturbed. These growth patterns often give information about an organism's relation to air and the surface activity of the organism.

Procedure

Period 1

1. Label each of 5 broth tubes with the name of one of the organisms to be inoculated. Label a sixth tube "control" and leave uninoculated.
2. Inoculate each of the tubes with one loopful of the appropriate stock cultures.
3. Incubate at 32°C for 48 hours.

Observations

Period 2

1. After incubation, compare inoculated tubes with the control by placing them against a bright background and also a dark

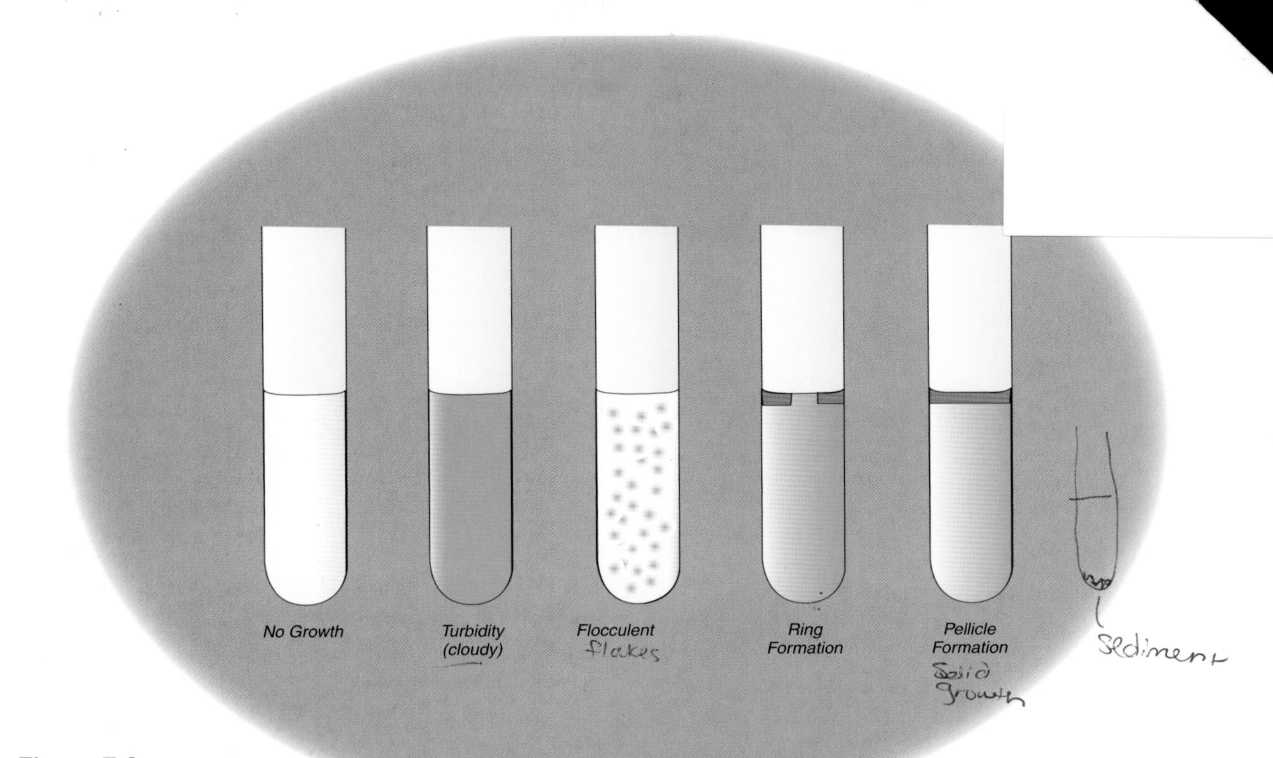

Figure 7-3

Broth cultural patterns

background. A readily available background is the ceiling fluorescent light fixture and the background ceiling panels. Hold the tube overhead across the junction of the light and the panel. Make observations using the diagrams in Figure 7-3.

2. Insert a sterile loop into each broth culture and slowly withdraw it. Note any tendency to form strings or slime. Sterilize the loop between each tube.
3. Complete the table in the laboratory report form.

B. Agar Slant Culture

Agar slants are prepared by melting a small amount of a solid medium (broth + 1.5% agar) in a tube and allowing it to solidify in a slanted position. Pure cultures of bacteria are streaked on the surface to serve as a **stock** culture for future use in inoculating media or to observe cultural characteristics on the slant. This last use will be described here. Note that the terms given in Figure 7-4 for agar slant culture patterns are not widely used in species or culture description any longer, because so many variables affect the pattern (e.g., inoculum size, amount of surface moisture, the medium, amount of agar included, the vigorousness of the culture's motility, and how steady the technician draws the loop up the slant). Other aspects such as pigmentation (chromogenesis) and consistency are still widely used, however.

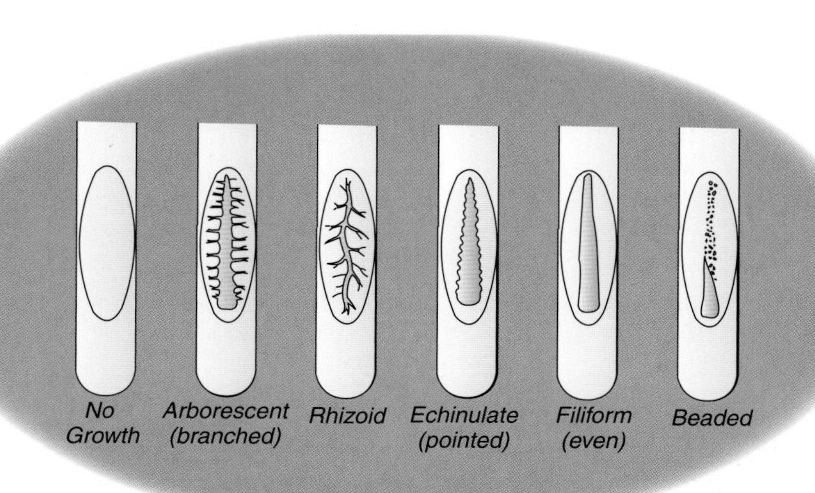

Figure 7-4

Agar slant cultural patterns

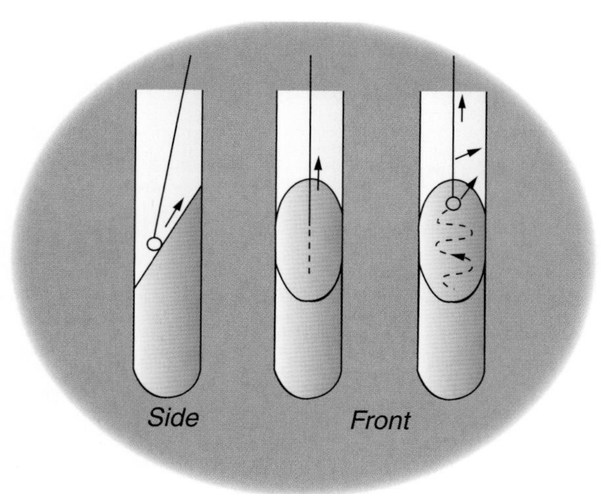

Figure 7-5

Holding the loop for making a slant culture

Procedure

Period 1

1. Label each of 5 agar slant tubes with the name of the organism to be inoculated. Label a sixth tube "control" and leave it uninoculated.
2. Using a loop, inoculate the slant labeled *Escherichia coli* with the stock culture provided. Streak the surface *in one straight line* from bottom to top while holding the open face of the loop vertically to the slant, as illustrated in Figure 7-5. Holding the open face of the loop horizontally will tear the agar. Rest the loop very lightly on the slant so as not to gouge. If the entire surface is to be covered, use the loop parallel to the agar surface and move it rapidly back and forth while drawing the loop slowly up the slant. Great care must be taken to avoid gouging the agar.
3. Repeat step 2 with each of the other cultures.
4. Incubate all tubes including the uninoculated control at 32°C for 48 hours.

Observations

Period 2

1. Examine the cultures using the diagrams in Figure 7-4 as a guide.
2. Touch the growth with a sterile loop and observe the consistency of the growth, whether it is brittle, soft, gummy, tough, slimy, buttery, or any other description you might use.
3. Complete the table in the laboratory report form.

8 Streak Plate

Objectives

The student will be able to:

1. prepare an uncontaminated plate for streaking.
2. streak and isolate the species within a mixed culture.
3. describe typical colonies with terms applied to colony morphology.

Adding a solidifying substance such as agar to a broth medium and then inoculating with bacterial cells traps individual cells in place. Instead of swimming or floating around when they multiply, as in a liquid medium, they are restricted to one place, forming a visible mass of growth called a **colony**. If the original cells are trapped some distance apart, each viable cell or clump of cells grows or develops into a separate, distinct colony. Because colonies differ in size, shape, texture, and color with different microorganisms, colony appearance is a useful aid in identification of species as well as a means of isolating the progeny of one cell as a **pure culture**.

You have thus far used solid media only in the confined area of a test tube. However, by increasing the surface area by placing the agar medium in a broad, circular **Petri plate**, new ways of handling microbial cultures become possible.

How to Prepare Agar Plates

After melting an agar medium, it is then cooled to 45°–47°C and 15–20 milliliters of the cooled, but still liquid, agar is poured into a sterile, covered Petri plate. Prior cooling prevents excess condensation of moisture in the Petri plates when the liquid agar solidifies.

Several **aseptic technique** precautions are necessary to prevent contamination. When you take the melted agar from the holding water bath, the **outside** of the container should be wiped free of excess water with a cloth or paper towel. Otherwise, the water will run down the outside of the container into the plate introducing contaminants from the water bath water. When removing the plug or cap to pour the agar, **flame** the mouth of the container to kill

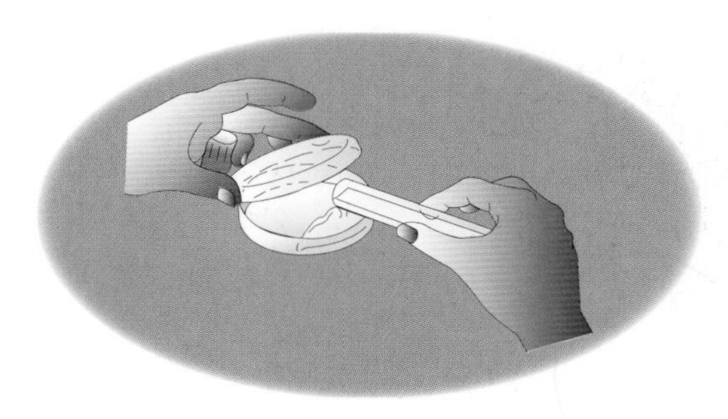

Figure 8-1

Using aseptic technique, lift the cover of the Petri plate high enough to insert the mouth of the tube to pour the melted medium into it. Do not touch the plate with the tube.

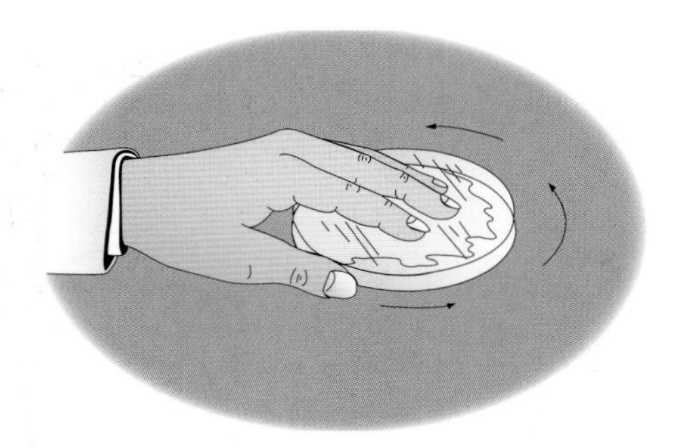

Figure 8-2

Rotate or slide the Petri plate in a figure-8 so that the medium covers the bottom. Do not move the plate again until the medium has solidified.

microorganisms on the outside lip and create an updraft of air from the hot glass. Do not overheat the lip. In pouring the agar from the container to the plate, raise the cover of the plate only on one side and just sufficiently to admit easily the mouth of the container. You must also take care not to touch the container to the Petri plate or its cover when pouring the agar.

How to Inoculate Agar Plates

You may now apply a microbial culture to the surface of the agar and spread it with a loop or bent needle. This is called **streaking** and a plate so prepared is called a **streak plate**. The purpose of the streak plate is to produce well-separated colonies of bacteria from concentrated suspensions of cells. Although there are a number of techniques you may use for streaking plates, only one is illustrated here. During the streaking, as cells are rubbed off the wire, the closely packed cells at the start of the streak form colonies that run together; however, as streaking continues, fewer and fewer cells remain on the loop or needle. As these are removed and grow on the agar surface, separate or **isolated colonies** develop. A few short hasty streaks will *not* produce isolated colonies. A good plate results from the progressive movements of the loop or needle, repeated many times.

The inoculating loop should be held at a slight angle to the agar surface with the open face *parallel* to the agar surface, as illustrated in Figure 8-3, rather than vertically to the surface which is almost guaranteed to produce gouges. Hold the handle so that the loop rests lightly on the agar surface. Slide the loop rapidly from side to side in arcs of 2-3 cm while drawing it toward you ***but without pressure*** (Figure 8-4). This reduces the likelihood of gouging the agar.

Rapidly growing or very prolific organisms may require flaming the loop *between* each streak or every other streak. If the organ-

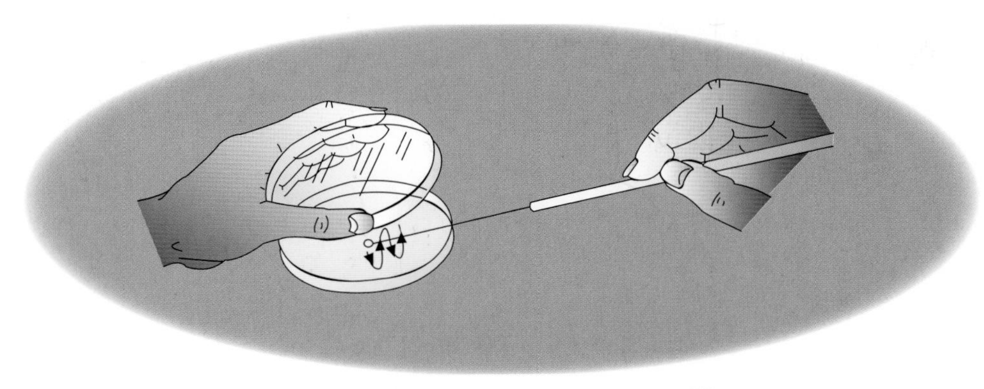

Figure 8-3

Proper handling of the cover of the Petri plate while preparing a streak plate. Hold the loop parallel to the agar to avoid gouging.

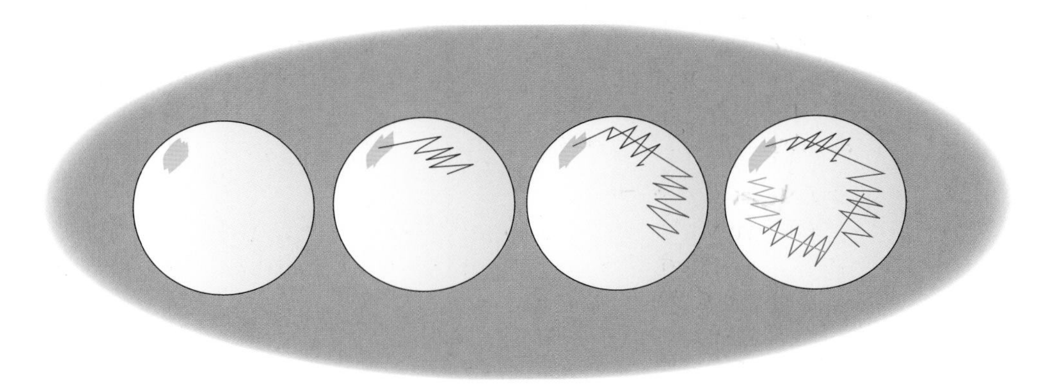

Figure 8-4

Steps in the preparation of a streak plate. The inoculating loop is sterilized in between each set of streaks.

ism grows poorly, then flaming between streaks can be omitted.

The best method of streaking a plate is to hold the plate in your hand using the thumb to rotate the cover upward allowing entry of the inoculating loop. If you cannot do this, place the plate on the bench and lift the cover just enough to allow entry of the loop as illustrated in Figure 8-3. One should *not place the cover on the bench* or hold the plate in the air. This increases the opportunity of contamination many-fold. To be reasonably sure of a pure culture, an isolated colony is streaked and restreaked four or five separate times.

How to incubate agar plates

Because of the high concentration of water in agar, some **water of condensation** forms in Petri plates during cooling and incubation. Moisture is likely to drip from the cover to the surface

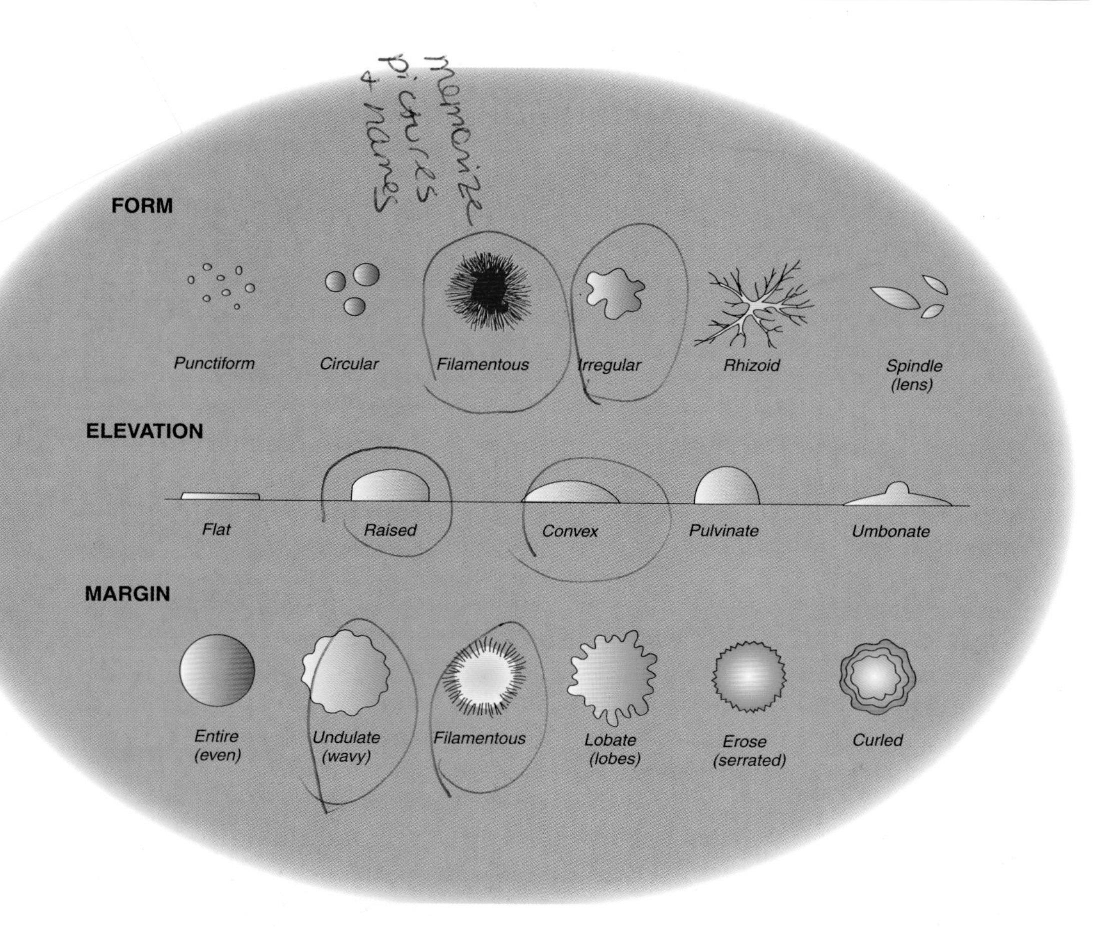

Figure 8-5

Cultural characteristics of isolated bacterial colonies. Colonies may be any color.

of the agar and spread out, resulting in a confluent mass of growth and ruining individual colony formation. To avoid this, Petri plates are routinely incubated **bottom-side up** (i.e., inverted). Note: There are exceptions to this rule with some procedures and media.

Materials

1. 18- to 24-hour mixed broth suspension of *Escherichia coli* and *Micrococcus luteus*
2. 3 nutrient agar deeps, sterile
3. 3 sterile Petri plates

Procedure

Period 1

1. Pour one Petri plate of nutrient agar using good aseptic technique. Allow it to solidify.
2. Flame the inoculating loop and allow it to cool for a few seconds.

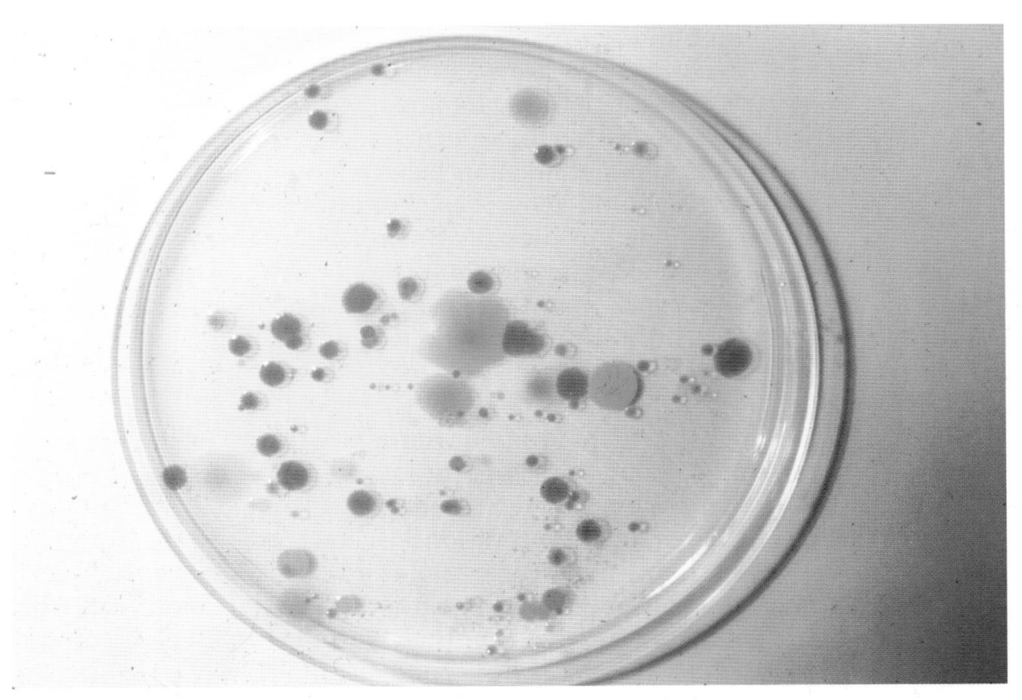

Figure 8-6

Halobacterial colonies from the Great Salt Lake, Utah, growing on 23% salt agar

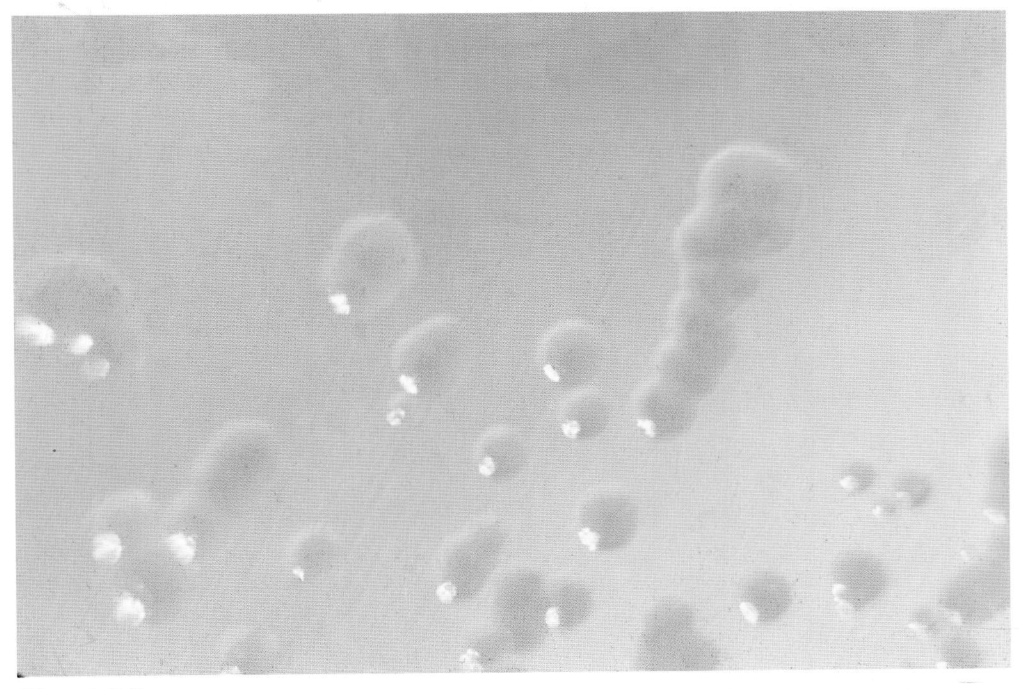

Figure 8-7

Yellow pigmented colonies of *Caulobacter* sp. from an aquatic sample

Take a loopful of a mixed broth suspension of *Escherichia coli* and *Micrococcus luteus* and streak back and forth over a small area near the side of the agar plate (Figure 8-4). Be careful not to cut the agar or contact the side of the plate. Lift the lid of the Petri plate only far enough to allow insertion of the loop.

3. Flame the loop and allow it to cool briefly. Pass the loop completely through the previously streaked area once and then back and forth in a restricted area as shown in Figure 8-4.
4. Repeat step No. 3 until there is no further room on the plate.
5. Incubate at 30°C for 24 hours.

Period 2

1. Have your instructor critique your streak technique on the plate before proceeding. This is an important part of developing good technique. (*Note:* The exercise may be terminated after this critique by making the observations called for in Period 3, step 1.)
2. Prepare two nutrient agar plates as you did in Period 1.
3. Select two **well isolated** and distinctly different colonies. Streak one for isolation on one agar plate and the other colony on the second plate. Touch only the center of the colony if possible.
4. Incubate both plates at 30°C for 24 hours.

Period 3

1. Make observations on isolated colonies as follows:
   Using the magnifying lens of the Quebec colony counter (Figure 8-8), make detailed drawings and record descriptions of growth for each species as called for in the table provided in the laboratory report form. Use the drawings in Figure 8-5 to aid in your description of the appearance of growth.

Figure 8-8

Quebec colony counter

9
Pour Plate

Objectives

The student will be able to:

1. perform a successful loop-dilution pour plate series.
2. explain why subsurface and surface colonies have different morphologies.
3. explain the principle of the loop-dilution pour plate method.

The technique of the **pour plate** gives you a second method for obtaining pure cultures from a mixture. It differs from the streak plate in that the agar medium is inoculated while it is still liquid but cooled to 45°C, and colonies develop throughout the medium, not just on the surface. In order to have isolated colonies on the plate, there is the problem of getting the proper concentration of cells into the plates poured, since there is no reliable method to predict how many viable cells may be in a given sample. If there are too many colonies, the plate is useless. The idea is to make successive **dilutions** of the cells and then on at least one of the poured plates colonies will be separated well enough to see them as isolated. Those plates that are too crowded or with too few (or none) are discarded. One limitation of this technique is that species with low numbers (rare) are not observed when other species are present in very high numbers. The rare species are overwhelmed by sheer numbers.

The **loop dilution** method as presented here will yield useful plates with many samples and is based on a roughly quantitative dilution of the original sample in an agar medium. A loop of 3 mm internal diameter, touching but not crossing the shank, will hold approximately .01 ml. A more quantitative procedure will be introduced in the following exercise and several later ones.

Subsurface colonies face a different environment from surface colonies. The agar gel restricts the growing colony physically and cells cannot spread evenly. More and more pressure is exerted by the growing colony until the agar gel actually breaks along a weak line and the cells grow into the crack. Subsurface colonies often assume a lens shape as a result. Sometimes the break will occur through the agar-plate interface at the bottom of the plate allowing cells to grow in a thin film of moisture between the agar and the plate surface. If this growth is visible, it assumes a flat, very thin, widely spreading colony sometimes referred to as a "ghost" colony. There may be many of these, and they make counting very difficult, as described in the next exercise.

Materials (per pair)

1. 18- to 24-hour mixed broth suspension of *Escherichia coli* and *Micrococcus luteus*
2. 3 nutrient agar deeps, sterile
3. 3 sterile Petri plates

Procedure

Period 1

Note: This exercise requires that you move very rapidly to inoculate, mix, and pour plates. Read the procedure thoroughly so you know ***exactly*** what to do for each step. If the medium solidifies in the tube, remelting **kills the organisms** and reinoculation is required.

1. Label 3 nutrient agar deeps No. 1 through No. 3 and place them into a boiling water bath for melting. Then allow them to cool to 45°C. While the tubes are cooling, label 3 Petri plates No. 1 through No. 3.

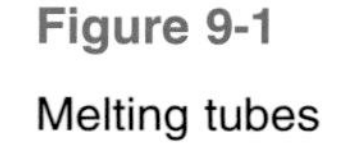

Figure 9-1

Melting tubes

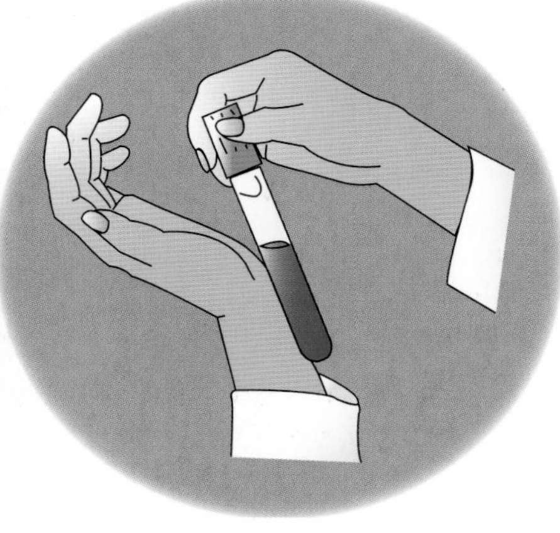

Figure 9-2

Testing agar temperature. Slightly warm to the touch, but not hot.

2. Prepare your loop carefully. It must be 3 mm in diameter and the loop closed so that the wire tip touches the wire shank. If not touching, a film does not form and the number of cells transferred is greatly reduced.

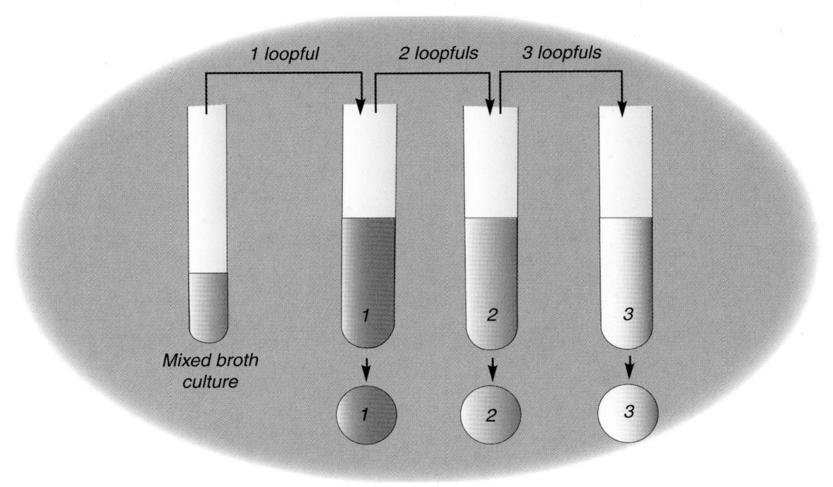

Figure 9-3

Steps in the loop-dilution pour plate method

3. Flame the inoculating loop and allow it to cool briefly. Transfer 1 loopful of the mixed culture of *Escherichia coli* and *Micrococcus luteus* to the first agar deep (Figure 9-3). Resterilize the inoculating loop. Mix by tapping the bottom of the tube rapidly with your forefinger and by rapidly rotating the tube between the palms of your hands.

Figure 9-4

Rotate the tube back and forth to ensure distribution of the inoculum

4. Flame the inoculating loop. Holding the first and second nutrient agar deeps, transfer 2 loopfuls of the suspension from tube No. 1 to tube No. 2. Reflame the loop. Mix tube No. 2 in the same manner as before.
5. Pour the contents of tube No. 1 into the Petri plate labeled No. 1. Swirl the plate in a figure-8, so that the agar covers the bottom of the plate.
6. Flame the inoculating loop. Holding the second and third nutrient agar deeps, transfer 3 loopfuls from tube No. 2 to tube No. 3. Reflame the loop. Mix tube No. 3 in the same manner as before.
7. Pour the contents of tube No. 2 into the Petri plate labeled No. 2 and swirl the agar gently.
8. Pour the contents of tube No. 3 into the third labeled Petri plate and treat as before.
9. After the agar has solidified, label the bottom of each plate with your name, the date, and the exercise number.
10. Incubate at 30°C for 24 hours.

Observations

Period 2

1. Examine each of the plates for isolated colonies using the magnifying lens of the Quebec colony counter. Make a sketch from a section of each of the three plates to illustrate the amount of growth, distribution of growth, and size of colonies. Note the appearance of subsurface colonies as compared with colonies growing on the agar surface (Figure 8-5).

10
Quantitative Dilution and Spectrophotometry

Objectives

The student will be able to:

1. handle pipets in an aseptic manner.
2. make a series of quantitative dilution plates to determine the number of bacteria per milliliter of a culture.
3. count the number of bacterial colonies on a plate according to the rules for plate counting.
4. calculate the number of bacteria per milliliter of sample from the dilution and the plate count.
5. use a spectrophotometer to determine the optical density of a bacterial culture and a series of its dilutions.
6. plot the relationship between optical density and bacterial numbers.
7. calculate the number of bacteria per milliliter of a sample from the plotted data given an arbitrary optical density.

Not only is it important to isolate microbes, it is also often necessary to know the number or mass of cells in a sample. To determine how many are present, a number of techniques are available: most probable number (**MPN**), direct microscopic counts, automated cell counters, plate counts, and many indirect methods. In this section, two of the most commonly used methods of determining numbers will be used: a direct method, the **viable plate count** (living cells grow), and an indirect method based on light absorbance. The latter method is often called "optical density" but is more accurately called absorbance. The absorbance is related to the count on the plates. Also introduced is the method of making **dilutions**, an integral part of many of the methods. The direct microscopic count and the MPN methods will be used in later exercises.

This exercise is designed to demonstrate a method used to enumerate viable bacteria in a liquid or a solid (e.g., food) and a

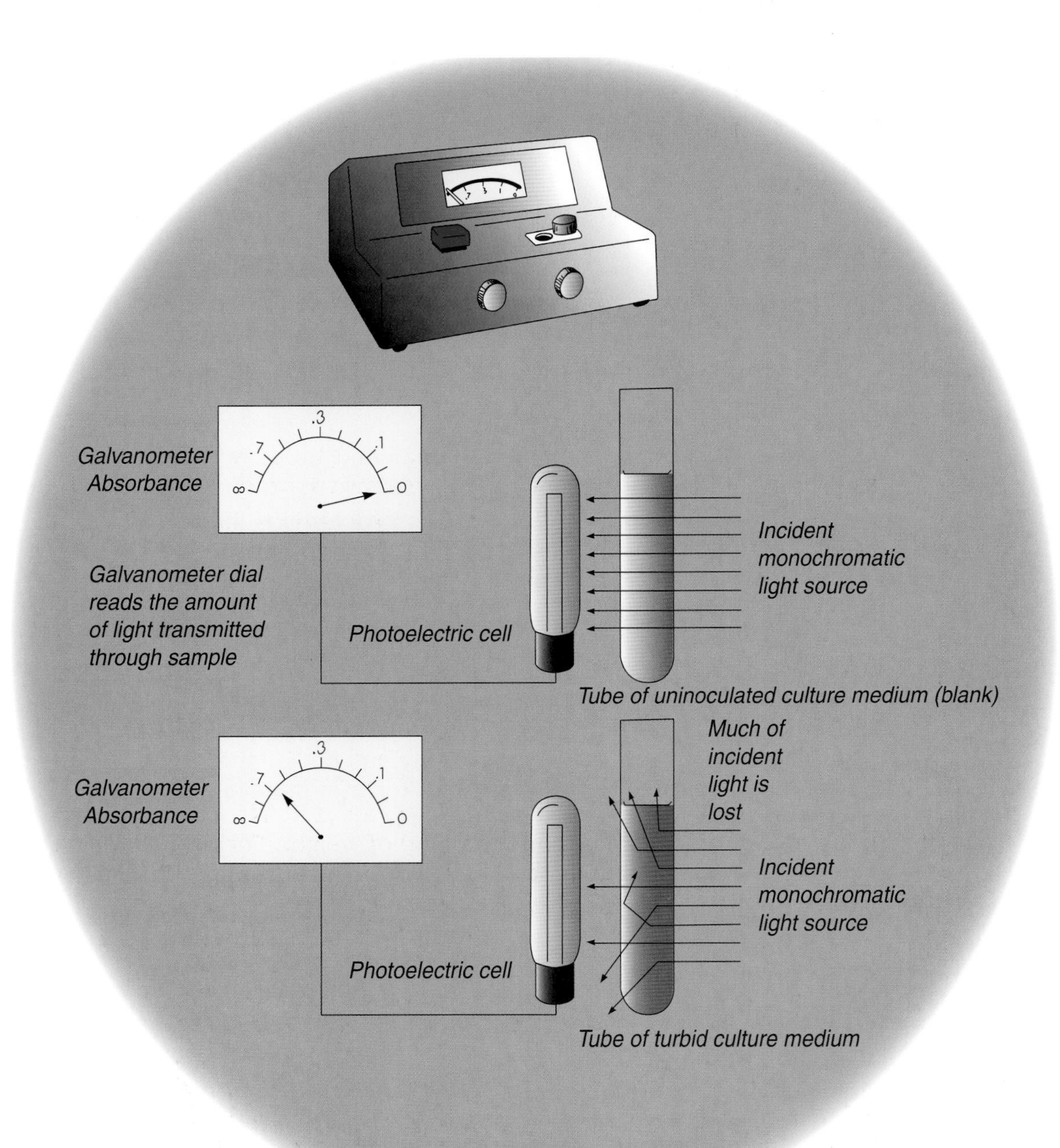

Figure 10-1

Principle of the optical method of measuring cell numbers

technique for measuring bacterial mass or density in a liquid by an optical method. The optical method can be made quantitative by relating **absorbance** (optical density or O.D.) to a number of bacteria per unit volume by the use of the first method.

The optical method of measuring growth depends on the interruption and scattering of a beam of light by the colloidal-size bacterial cells (Figure 10-1). The amount of light lost or scattered is inversely proportional to the cell concentration or directly proportional to the **absorbance** (optical density). The light loss can be determined by measuring the amount scattered or reflected (nephelometry) or the amount of light transmitted (turbidimetry). Scales on galvanometers are measured in % transmission (%T) and/or absorbance. The two scales are related as follows:

$$\text{absorbance} = \log 100 - \log \%T$$

The technique assumes that the glassware used is optically matched (i.e., all tubes have the same optical properties). This exercise is best done with optically matched tubes, which may not be available. The use of regular glassware produces some variation, and the curve plotted may deviate somewhat from linearity as a result. The laboratory instructor will provide more detailed information on the instrument used for this exercise.

Unless a given absorbance reading can be quantitatively related to a number of bacteria per milliliter, the density scale must be in arbitrary units (e.g., McFarland turbidity standards). In order to quantify the absorbance reading, a quantitative enumeration must be made of the number of bacteria in a unit volume of the suspension used. The **quantitative dilution** method used in this exercise is a variation of the pour plate technique of Exercise 9.

A unit of **sample** (the material being analyzed) is usually one milliliter (ml) of volume or one gram (g) of weight. For purposes of quantitative dilution, one ml of water is considered to be equivalent to one g. One ml (or g) is added to a sterile known volume of water or buffer. While any volume of dilution water can be used, common practice calls for 9 or 99 ml volumes for ease of calculation. **Serial dilutions** are made (Figure 10-2) and samples plated out from each dilution. After incubation, colonies are counted on plates with 30–300 colonies. Plates with colony numbers outside this range are discarded unless no plates with 30–300 are encountered. The number of colonies is multiplied by the **reciprocal of the dilution factor** (i.e., the total dilution representing the plate counted), which gives the number of bacteria per ml or g of original sample. Dilutions are generally expressed as negative exponents (e.g., 10^{-4} rather than 1/10,000).

One important aspect of this exercise is the use of pipets. Pipets are cylinders of glass or plastic calibrated to hold a certain volume of liquid. Together with a pipet aid, the total volume or a discrete fraction of the volume can be released for making dilutions or placing a volume in a plate or tube. Most pipets are of the transfer or bacteriological type, which requires that the last drop be expelled by the aid in order to get the full volume. These usually have two rings around the handling end. Some pipets are calibrated to the tip (no rings). Look carefully so you will know what kind of pipet you have. Ask your instructor if in doubt. You should **NEVER** attempt to use your mouth in place of a pipet aid. Pipets are supplied in a sterile condition in a container of metal or a plastic bag. The container must be opened or the plastic cut so that pipets can be removed by handling the **untapered** end of the pipet without contamination from external sources. Pipets should be handled only at the extreme end and at no time should you touch any part that enters a tube, plate, or dilution blank. Pipets should be taken **ONLY AS USED, ONE AT A TIME**. Pipets should be seen in the laboratory **ONLY** (a) in the **original** container, (b) **ONE** in your hand, or (c) in the **discard** container after use. They should **NEVER** be removed in multiples and **NEVER** placed on the bench top before or after use. Furthermore, they should **NEVER** be returned for **ANY REASON** to the original container once removed. Pipets should be discarded **only** in containers specifically for that purpose.

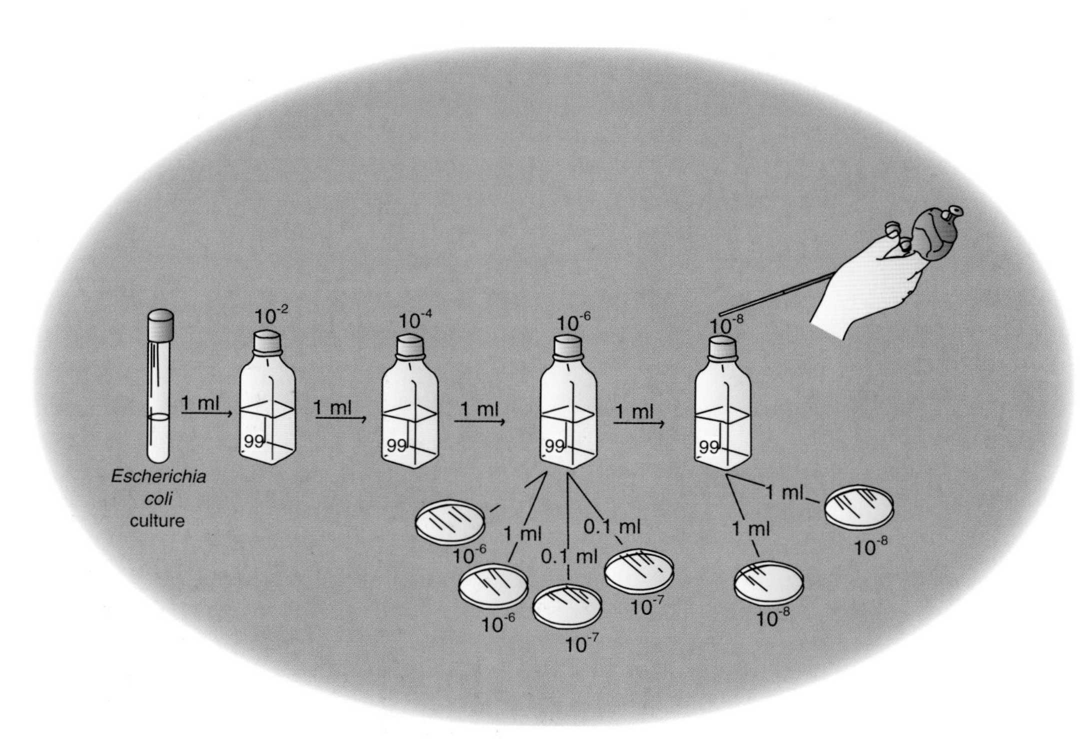

Figure 10-2

Dilution plating procedure

Materials (per pair)

1. 18- to 24-hour broth culture of *Escherichia coli*
2. 5 13 mm screw-capped tubes each containing 2 ml nutrient broth, sterile
3. 6 nutrient agar deeps, sterile
4. 6 sterile Petri plates
5. 1 sterile 5 ml (or 2 ml) pipet and pipet aid (a 1 ml pipet can be used)
6. 5 sterile 1 ml pipets calibrated in tenths
7. 4 sterile 99 ml dilution blanks
8. Spectronic 20 (or other spectrophotometer) with test tube holder
9. Quebec colony counter
10. Hand tally counter
11. Discard container for pipets

Procedure

Period 1

A. Determination of Bacterial Count Per Unit Volume of Culture

In order to know how many bacteria per ml the absorbance represents, you must first determine the number of bacteria per unit volume of the culture used. Because aseptic technique is essential, this part is done **FIRST**.

1. You will be provided with one tube of an 18- to 24-hour-old culture of *Escherichia coli.*
2. Prepare a serial dilution of the culture as follows (Figure 10-2):
   a. Melt 6 nutrient agar deeps and cool to 45°C. Keep them at this temperature until ready for use.
   b. With a 1 ml pipet and aid, transfer 1 ml from the culture tube to the first 99 ml dilution blank marked 10^{-2}. **Discard** the pipet. Shake the dilution blank 25 times in a 1-foot arc in 7 seconds (Figure 10-3). *Save the original culture* for part B.

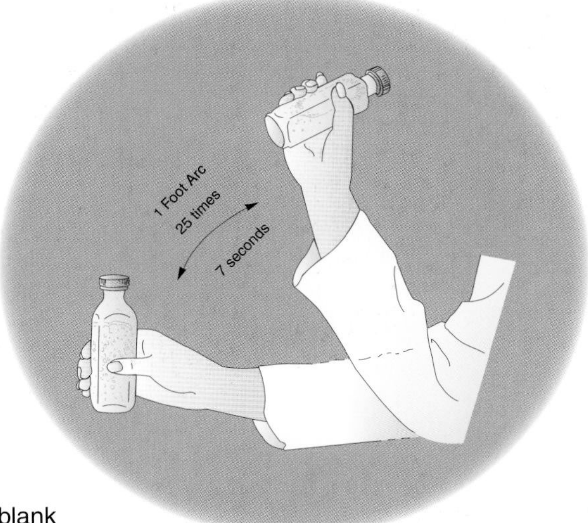

Figure 10-3

Procedure for mixing a sample in a dilution blank

c. With a new sterile pipet, transfer 1 ml from the first dilution blank to a second marked 10^{-4}. **Discard** the pipet. Shake the dilution blank as before.
d. With a new sterile pipet, transfer 1 ml from the second dilution blank to a third marked 10^{-6}. **Discard** the pipet. Shake the dilution blank as before.
e. With a new sterile pipet, transfer 1 ml from the third dilution blank to a fourth marked 10^{-8}. **HOLD** the pipet. Shake the dilution blank as before.
f. With the same pipet used in step 2.e, pipet 1 ml from the third dilution blank (10^{-6}) to one plate marked 10^{-6} and 1 ml to a second plate marked 10^{-6}. Then pipet 0.1 ml to one plate marked 10^{-7} and 0.1 ml to a second plate marked 10^{-7}. **Discard** the pipet.
g. With a new sterile pipet, transfer 1 ml from the fourth dilution bank (10^{-8}) to a plate marked 10^{-8} and 1 ml to a second plate marked 10^{-8}. **Discard** the pipet.
h. Pour one nutrient agar deep into each Petri plate and rotate gently in a figure 8 to mix the agar and the sample. Move slowly, being careful **NOT** to slop agar on the cover.
i. Allow the agar to solidify.
j. Invert and incubate the plate at 37°C for 24–48 hours.

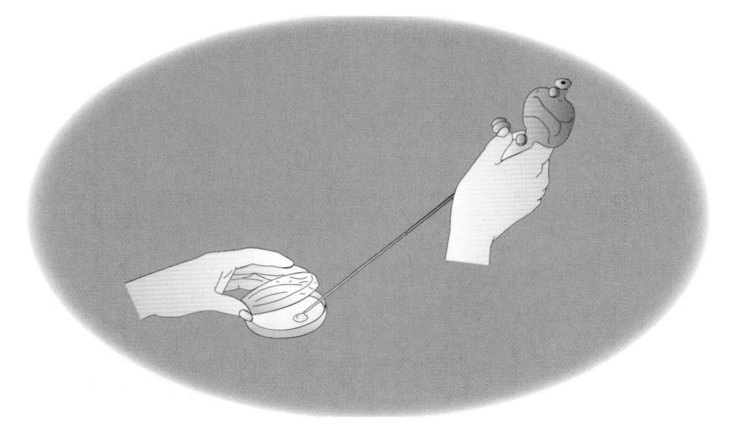

Figure 10-4

Technique for holding a pipet and a Petri plate cover

B. Absorbance vs. Bacterial Count

Now you must determine the absorbance of a series of dilutions of the original culture and plot a curve. By measuring the absorbance of an unknown culture of the same organism in the same medium you can determine the number of cells per unit volume of the unknown from the curve.

1. With the remainder of the original tube of *Escherichia coli* (after step 2 above) proceed as follows in Figure 10-5:

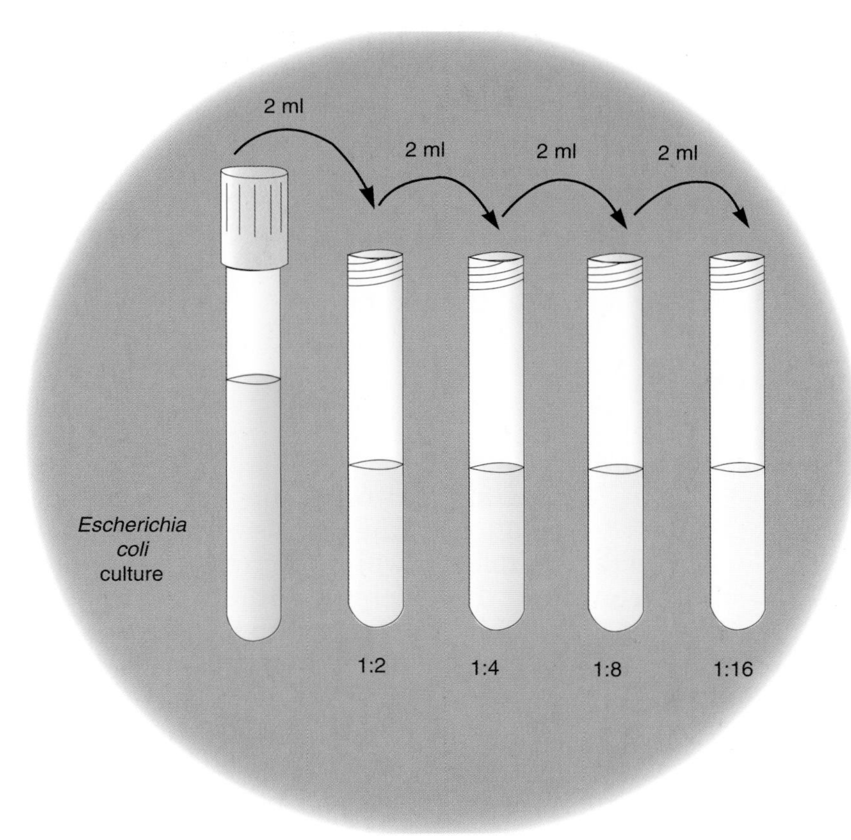

Figure 10-5

Procedure for preparing serial dilutions for optical density determinations

a. Obtain 5 tubes of nutrient broth, each containing 2 ml, and make serial 1:2 dilutions (use the **same** 2 ml (or 5 ml) pipet and aid throughout):
   i. Transfer 2 ml of the original *Escherichia coli* culture to one of the sterile broth tubes. Mix well. Mark the tube 1:2.
   ii. Transfer 2 ml of the 1:2 tube to another broth tube. Mix well. Mark it 1:4.
   iii. Transfer 2 ml of the 1:4 tube to another broth tube. Mix well. Mark it 1:8.
   iv. Transfer 2 ml of the 1:8 tube to another broth tube. Mix well. Mark it 1:16.
   v. Do **NOT** transfer anything to the last broth tube but mark it "control" or "blank".
b. Using the Spectronic 20 (or other instrument) provided, set the wavelength at 525 nm (refer to Figure 10-1).
   i. Adjust the galvanometer to the 0% T (or absorbance) mark with the left front knob.
   ii. Insert the "control" broth tube into the holder and replace the cap. Adjust the galvanometer to 0 absorb-

ance using the right front knob. This measures the absorbance of an uninoculated tube of the same medium and sets the absorbance to zero. Any change in absorbance in other tubes is then due to the turbidity of the cells.

iii. Beginning with the original *E. coli* tube, insert each dilution tube prepared in B.1.a. into the holder, replace the cap, and record the absorbance for each tube on the report form.

Observations

Period 2

A. Rules for Counting Colonies

1. Following incubation and using the Quebec colony counter (Figure 10-6), count all of the colonies on each of the duplicate plates at the one dilution with all plates having 30–300 colonies and determine the average. Otherwise go to step 2. Occasionally colonies will break through the agar surface next to the plastic or glass bottom, allowing cells to grow in a thin film of moisture there. These result in thin, almost transparent, colonies sometimes referred to as "ghost" colonies. These should all be lumped together as a single colony unless there is a significant distance (at least 3 cm) between them. Individual plates may be unusable for various reasons, in which case the following codes and comments may be used:

Figure 10-6

Quebec colony counter

Spr = spreader (*if more than half of the plate is covered with more or less uniform growth, do not use for counting*)
NC = no colonies (see 2e below for additional direction)
LA = laboratory accident
GI = growth inhibitor (only when true growth inhibition is observed, i.e., lower dilutions have lower counts than higher ones. Beware: this is more likely to be an error in dilution - LA instead)
TNTC = too numerous to count (do not use unless the plate is truly uncountable or with confluent growth at the highest dilution)

Now proceed to step 3.

2. In case the initial condition of step No. 1 cannot be met (i.e., **all** the plates at **one** dilution having 30–300 colonies), select one of the following procedures:
   a. If one of the plates is within the 30–300 range and the other outside, count both and average. Proceed to step 3.
   b. If plates from **two consecutive** decimal dilutions yield 30–300 colonies each (rare), compute the count per ml for **each dilution** separately. If the **higher** calculated count is more than **two times the lower count**, then use the lower count as the result. If the **higher** count is **less than** two times the lower count, then average the counts for **both** dilutions and proceed to step 3.
   c. If **no** plate has 30–300 colonies but one or more has **more than** 300, use the dilution with the counts **closest** to 300. Use the colony counter grid (Figure 10-6) and proceed according to one of the following:
      i If there are **fewer than 10** colonies per square cm (the larger squares), count the colonies in 11 squares (for plastic plates; 13 if glass) representative of the distribution of the colonies. Multiply this total by 5 to give the total number of colonies on the plate. (The plastic plate is 56 cm^2; 65 cm^2 if glass.) Average the plates. Counts determined in this manner must be reported as "Estimated". Proceed to step 3.
      ii If there are **more than 10** colonies per square cm (the larger squares), count the colonies in 4 representative squares, average and multiply by 56 (if plastic; 65 if glass) to give the number of colonies on the plate. Average the plates. Counts determined in this manner must be reported as "Estimated". Proceed to step 3.
      iii If there are **more than 100** colonies per square cm (the larger squares), assume the count to be >6500 on the plate and report as "Estimated >6500". Do **NOT** report as TNTC.

Proceed to step 3.

d. If **fewer than 30** colonies are found on each plate, record the actual number and average. Report such counts as "Estimated". Proceed to step 3.

e. If there are **no colonies** on **any** dilution plates, report the count as **"Estimated less than the reciprocal of the dilution factor".** If no colonies were found on the 10^{-2} plate, then the count would be "Estimated <100".

Proceed to step 3.

3. In all cases, multiply the average count (or single count if one plate used) times the reciprocal of the total dilution of the plate at which the count was made. The product is the count of bacteria per ml of original sample (i.e., the original culture used). After calculating the count per ml, treat the result as follows:
   a. The final count must be rounded to two significant figures to avoid fictitious accuracy. To do this, each digit is rounded to the next highest number if the one to the right is 5 or above or to zero (0) if 0 to 4. For example 146 becomes 150; 144 becomes 140.
   b. Scientific notation is the usual method of reporting unless the count is below 100. In the example above, 150 would be reported as 1.5×10^2/ml or g.
   c. Add the word "Estimated", if required.

B. Relationship of Culture Count to Absorbance

1. From step A.3, enter the count, the dilution counted, and the final count per ml of the **original** *E. coli* culture on the report form in the space provided under part A. Place this same count per ml in part B in the column under "Bacteria per ml" opposite "undiluted" *and* also under the graph under the space marked "none" (i.e., undiluted culture).
2. Knowing that you made serial two-fold dilutions, calculate the bacterial count per ml for each of the dilutions. For example 1:2 would be one half the undiluted count, 1:4 would be one fourth the undiluted count, etc. Enter these values in the table and under the graph.
3. Now, plot the absorbance versus the dilution by placing a mark at the appropriate spot on the graph and connecting the resulting points with a ruler. Ideally, all the points should fall on a straight line, but this does not always happen, especially if optically corrected glassware is not used. Use your straight edge to make a line of best fit (i.e., a straight line that comes as close as possible to all the points).

UNIT III

Staining Techniques

Although the microscope revealed to van Leeuwenhoek a new world of "animalcules", this world remained essentially invisible until staining techniques were developed to distinguish the microbes from their environment, as well as from each other. Unstained microorganisms are nearly transparent when observed by light microscopy and as a result are difficult to see. Although modern microscopy has helped through the development of phase contrast and interference microscopes, among others, staining of cells to make them more visible still remains the standard tool. In addition to making cells more visible, special stains have been developed to detect special structures or chemicals in the cells. Staining techniques are available (1) to permit easier visualization microscopically by providing contrast between microorganisms and their background, (2) to allow a more detailed examination of cells by utilizing higher magnifications, (3) to identify internal structures of cells, and (4) to provide a means of differentiation of microbial groups.

Most of the commonly used stains are benzene derivatives called **aniline** dyes from coal tar, or **synthetic dyes**. Each dye molecule has two functional chemical groups with properties that respond to the wavelengths of light resulting in a color. The visual detection of color from these dye compounds is due to the **chromophore group**, a chemical structure consisting of unsaturated bonds that absorb specific wavelengths of light. The **auxochrome group** is an ionizing chemical structure that gives the molecule increased solubility and salt-forming characteristics and the ability to react chemically with an ionized substrate. Dyes are classified as **anionic** (acidic) or **cationic** (basic) depending upon whether the chromophore is an anion (negatively charged) or a cation (positively charged). They are sometimes imprecisely referred to as acidic or basic dyes, terms that are really misnomers.

Cationic dyes (e.g., methylene blue, crystal violet, carbolfuschin) will react with chemical groups in the substrate that ionize to produce a negative charge. Anionic dyes (e.g., eosin, nigrosine) will react with positively charged groups. Since microbes possess a variety of both types of groups on the exterior or interior of the cell, they have the ability to combine with cationic and anionic dyes.

For practical purposes, microbial stains can be divided into three groups: simple, differential, and special/structural stains. **Simple stains** (e.g., methylene blue) are single dyes used primarily to help visualize the normally transparent cells and do not differentiate between kinds of organisms. The basis for simple staining is the fact that the cells differ chemically from their environment and can, therefore, be stained to contrast with that environment. **Differential stains** depend on the chemical and physical differences between different kinds of cells. The Gram and Ziehl-Neelsen stains are examples used to differentiate bacteria. **Special** or **structural stains** are used to stain particular physical or chemical structures (e.g., flagella, endospores, inclusion granules, DNA). Stains of this last category are used commonly as aids in identifying bacteria in particular.

In this unit, students will have an opportunity to use a simple stain, two differential stains, and a structural stain, as well as learn how to make a smear, a critical step in any staining procedure.

11
Smear Preparation and Simple Staining

Objectives

The student will be able to:

1. prepare a smear of proper density from a slant culture.
2. fix a smear with heat.
3. use a simple stain on a smear from teeth and gums.
4. describe the morphology of mouth and saliva organisms.

Unstained microbial cells are nearly transparent when observed by light microscopy and hence are difficult to see. Various staining techniques are available to permit easier visualization microscopically, a more detailed examination of cells, observation of internal cellular components, and a differentiation of cell types.

In order to stain microorganisms, a thin layer of cells called a **smear** must be made first. This is a simple process of spreading an aqueous suspension of cells on a glass slide and allowing it to air dry. This is followed by **fixation** (causing the cells to adhere to the slide) and the application of the staining solutions. Stains are generally made on smears from colonies or slants, since the mass of cells is very great. The main problem with this is getting too many cells in the smear. Making smears of the proper density can be learned only by experience. Smears can be made from broth cultures but several problems arise. The number of cells is usually very small, making them hard to find on the slide, and nutrient carryover causes difficulty in fixing the cells to the slide, allowing them to wash off more easily. Organic matter and salts also may interfere with the staining process. Smears can also be made from other materials providing there are enough cells present to see under the microscope. At least 500,000 cells per ml are necessary for oil immersion. In this exercise, saliva and teeth scrapings are stained as well as pure culture smears. Tap water is often used to make smears since it has too few bacteria (<10,000 per ml) to interfere with observations.

This exercise has two objectives: one, to practice making a smear of the proper density for staining; and two, to use a simple stain on it and/or a smear of material from around the teeth.

Materials

1. 18- to 24-hour slant culture of *Escherichia coli*
2. 3 microscope slides
3. Cleanser
4. 95% ethyl alcohol
5. 13 mm clean test tube (need not be sterile)
6. Loeffler's methylene blue stain
7. Marking pen
8. Clothespin or slide holder
9. Staining racks
10. Paper towels

Procedure

Period 1

A. Cleaning of Microscope Slides

Clean, grease-free slides are essential for obtaining good stained preparations. Before beginning any staining procedure, slides should be cleaned as follows, unless new, unused, precleaned slides are supplied:

1. Wet the tip of your index finger and rub it on some abrasive cleanser (e.g., Bon Ami, Comet, Ajax, etc.).
2. Spread the paste formed over both surfaces of the slide.
3. Wash the slide thoroughly with running water.
4. Apply several drops of 95% ethyl alcohol to the slide and allow it to air dry.
5. Flame the "up" side of the slide for a moment in the Bunsen burner. Allow it to cool.

B. Smear Preparation from a Solid Culture

1. Label the slide with your name or the procedure and then turn the slide upside down so the label is on the bottom.
2. Place 2 to 3 loopfuls of tap water on the top side of the slide.
3. Flame the inoculating loop and cool for a few seconds.
4. Touch the growth on the slant lightly with *one side* of the loop (Figure 11-1).
5. Very briefly touch the side of the loop with the cells to the drop and move it back and forth *once.*
6. Spread the cells to an area the size of a dime using the *other side* of the loop where no cells were present.
7. Resterilize the inoculating loop.
8. Allow the smear to air dry for at least 30 minutes. When dry, the

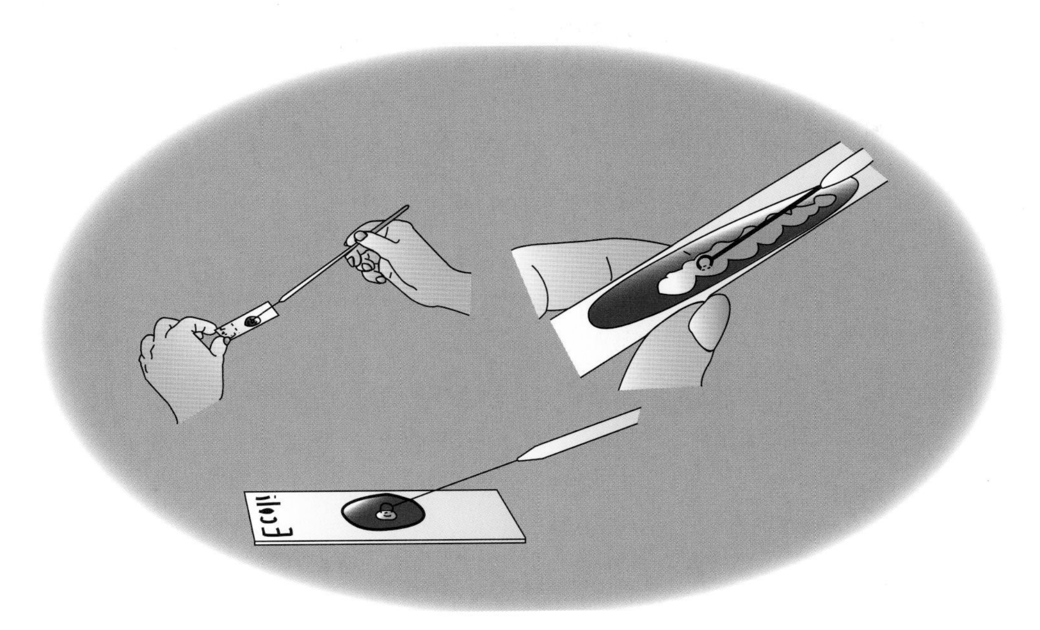

Figure 11-1

Preparation of a smear from a slant culture. Note that the label is on the opposite side from the smear.

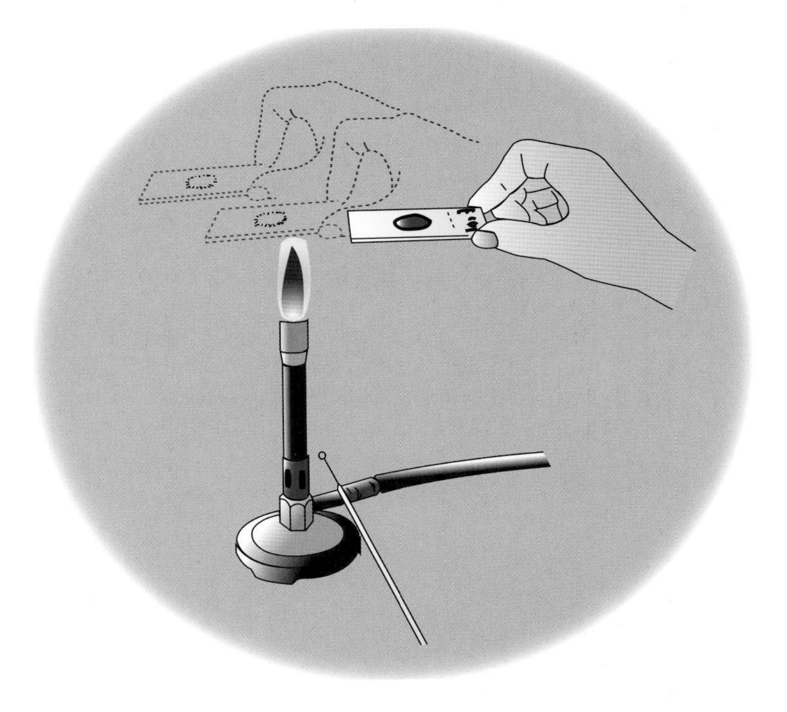

Figure 11-2

Heat fixation of an air-dried specimen

smear may appear faintly cloudy or very little may be seen, depending upon the amount of growth on the slant. It should be only **faintly** cloudy or turbid. It should ***NOT*** look like a drop of milk.

9. The smear must be "heat fixed" to be certain that the cells will adhere to the slide and not be washed off during staining procedures. This is accomplished by passing the slide (smear side up) back and forth through the flame of the Bunsen burner slowly three times (Figure 11-2). Do *not* overheat the slide. You should be able to touch the underside of the slide comfortably to the back of your hand.
10. Have your instructor critique the smear after drying. This is important, since first smears are often too heavy for good staining.
11. If your instructor so indicates, this smear may be saved for use in Exercise 12.

C. Smear Preparation from Teeth

1. Force a little saliva rapidly back and forth across your front teeth and collect it in a test tube.
2. With a sterile inoculating loop, transfer 2 loopfuls to a clean slide and spread to form a thin film.
3. Resterilize the inoculating loop.
4. Allow the smear to air dry.
5. Heat fix as in step No. 9 of Part B.

D. Staining

1. Place the slide from Part A and/or B on a staining rack and flood with Loeffler's methylene blue for 30–60 seconds.
2. Rinse *quickly* with water and blot dry. Too long in the water removes most of the stain.

Observations

Period 2

1. Examine under oil immersion and make drawings of what you see. Look for very small cells of various shapes from both sources and from the teeth smear look also for large white blood cells (lymphocytes).

12
Gram Stain

Objectives

The student will be able to:

1. prepare a Gram stain with consistent results.
2. know the Gram reaction and morphology of the organisms in this exercise.
3. recognize and describe the purpose and results of each step of the procedure.
4. list the reagents used and the timing of application.

Simple staining depends upon the fact that bacterial cells can be stained to contrast with their environment. Microorganisms also differ from one another chemically and physically and, therefore, may react differently to a given staining procedure. This is the basic principle of **differential staining**, a method of distinguishing between types of bacteria.

One differential stain discovered very early in the development of microbiology by Christian Gram is called the Gram stain in his honor. The Gram differentiation is based upon the application of a series of four chemical reagents: a **primary dye**, a **mordant**, a **decolorizer**, and a **counterstain**. The purpose of the primary dye, *crystal violet*, is to impart a purple or blue color to **all** organisms regardless of their ultimately designated Gram reaction. This is followed by the application of *Gram's iodine*, which acts as a mordant by forming a dye-iodine complex. The decolorizing *acetone-alcohol solution* extracts the complex from certain cells more readily than others. A *safranin* counterstain is applied in order to see those organisms previously decolorized by removal of the complex. Those organisms retaining the complex are **Gram-positive** (purple or blue), while those losing it are **Gram-negative** (pink or red). The ability to retain the dye-iodine complex or to lose it depends on the nature of the cell wall structure. Bear in mind that excessive application of the decolorizer will remove the dye from any cell, so the timing of the steps is critical to the end result. Although most cells, including the cyanobacteria and eukaryotic cells, will stain with the primary dye, they are not considered Gram-reactive because the decolorization step does not usually result in decolorized cells. The terms Gram-positive and Gram-negative are used only with the bacteria.

Materials

1. 18- to 24-hour slant cultures of *Escherichia coli*, *Bacillus cereus*, and *Staphylococcus aureus*
2. Gram's crystal violet (with ammonium oxalate added)
3. Gram's iodine (Hucker modification)
4. Acetone-alcohol
5. Gram's safranin
6. Slide holder or clothespin
7. Marking pen
8. Microscope slides
9. Staining rack or pan
10. Paper towels

Procedure

Period 1

1. Prepare smears of *Escherichia coli*, *Bacillus cereus*, and *Staphylococcus aureus*. Air dry and heat fix. All 3 smears are best placed on a single slide as illustrated (Figure 12-1), but separate slides can be used. Be sure to label the smears carefully so that you can distinguish one organism from another. Smears on a single slide are most suitable, since all smears are treated the same. This will be important later in controlling the staining procedure when an unknown (designated *other* in Figure 12-1) organism is used.

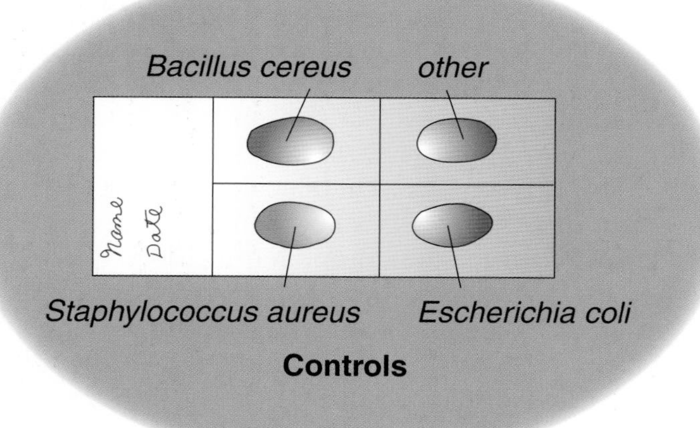

Figure 12-1

Preparation of a smear to be Gram stained

2. Apply Gram's crystal violet for 1–2 minutes.
3. Wash off the excess stain by holding the slide briefly under a stream of tap water. Do not tilt the slide until it is under the water. Shake off the excess water gently. Do not let the slide become dry before the next step.
4. Flood the slide with Gram's iodine and allow it to react for 1 minute or longer.
5. Rinse as in step No. 3. Shake the excess water off or blot lightly, but ***not* to dryness**. Becoming too dry is a common cause of failing to stain properly.
6. Holding the slide at an angle, carefully add the decolorizing solution one drop at a time. As soon as color stops coming off the slide, after about 8–10 seconds, rinse with water to stop the decolorizing action, and shake off the excess water gently.
7. Flood the slide with Gram's safranin and allow it to react for 30–60 seconds.

8. Drain the excess stain from the slide and wash it with tap water.
9. Carefully blot the stained slide using a paper towel. **Do *not* rub**.

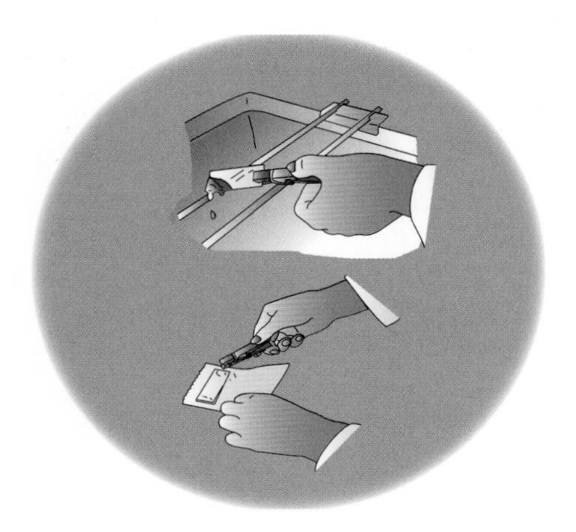

Observations

Period 1

1. Examine with the oil immersion objective. Make a representative drawing of each organism, noting Gram reaction and morphology.

   These organisms are used for several reasons. *Escherichia coli* is invariably Gram-negative, while *Staphylococcus aureus* is always Gram-positive unless grossly over-decolorized. *Bacillus cereus* represents a group that is called Gram-variable (i.e., it decolorizes so easily that a smear may actually have both Gram-positive and Gram-negative cells at the same time). If the *E. coli* and *S. aureus* cells are properly stained, you have done your stain properly and the *B. cereus* is then done properly also regardless of what reaction you see. Gram-variable organisms are usually considered to be Gram-positive. The *E. coli* and the *S. aureus* serve as staining controls for proper technique. In addition, *B. cereus* also may form internal structures called endospores which do not stain, thus appearing as a clear, unstained part of the cell.

   Other notes:

   - Gram staining should be done only on cultures of 12–24 hours' age.
   - After 24 hours, Gram-positive cultures often become Gram-negative or Gram-variable.
   - Gram-negative cells almost never become Gram-positive.
   - Gram stains from surface growth are preferable to broth cultures, because there are more cells in the smear and the broth ingredients often interfere with the staining process.

13
Acid-Fast Stain

Objectives

The student will be able to:

1. perform an acid-fast stain.
2. list the reagents used and the method of application for each step of the procedure.
3. describe the purpose and result of applying each reagent.
4. name the genera in which the acid-fast property is found.
5. name two human diseases caused by acid-fast organisms.
6. describe the morphology and chemical characteristics of acid-fast organisms.

Acid-fastness is a characteristic limited to the members of the genus *Mycobacterium* and a few *Nocardia* species. These bacteria possess a high concentration of waxy lipids in the form of mycolic acids in the cell wall. This layer of lipids imparts an almost impenetrable coating, making the cell most difficult to stain; therefore, special procedures must be used. In the **Ziehl-Neelsen** method, *carbolfuchsin* (red) is utilized as the primary stain. Because of the waxy, impermeable cell wall, heat is required to drive the dye into the cell. The decolorizer, consisting of a mixture of *acid and alcohol*, removes the red of the carbolfuchsin readily from all organisms except the acid-fast bacilli. Counterstaining with *methylene blue* colors the non-acid-fast organisms blue, while the acid-fast organisms appear red. As with many procedures, the dye of acid-fast organisms can be removed if over-decolorized. Care should be taken to avoid this. The waxy coating of the acid-fast bacilli presents an additional problem in that cells tend to adhere together and form clumps in smears rather than being spread uniformly. One must search the slide carefully to make sure they are not overlooked. The student needs to be very careful **NOT** to mistake the Gram stain decolorizer for the acid-alcohol decolorizer or vice versa. They look identical, and the acid-alcohol destroys the Gram-positive reaction on contact. The acid-fast bacteria are medically important as agents of tuberculosis and leprosy and a few other infections.

Materials

1. 18- to 24-hour slant culture of *Staphylococcus aureus*
2. 72-hour slant culture of *Mycobacterium smegmatis*
3. Ziehl-Neelsen carbolfuchsin
4. Acid-alcohol (acid-ethanol)
5. Loeffler's methylene blue
6. Slide holder or clothespin
7. Marking pencil
8. Microscope slides
9. Staining rack or pan
10. Paper towels

Procedure

Period 1

1. Prepare a slide as illustrated in Figure 13-1. Make a smear of *Staphylococcus aureus* on one end of a slide and *Mycobacterium smegmatis* on the other. Between them, prepare a mixture of the two. Make sure you **do not** go back into each culture tube at any time without re-sterilizing the loop. Air dry and heat fix the smears.
2. Flood the smears with the Ziehl-Neelsen carbolfuchsin stain and allow it to stand for about 1 minute.
3. Heat the preparation to steaming by inverting your Bunsen burner and passing the flame over the stain, or under the slide, moving back and forth the width of the stain as in Figure 13-2.
4. Remove the burner when you see steam rising from the stain. When the steam stops rising, pass the flame over or under the stain again as necessary to keep the smear just at steaming. Steam for 5 minutes. Do not boil nor allow the smear to dry. Add the carbolfuchsin as needed to prevent loss of the stain by evaporation.
5. Allow the slide to cool to prevent breaking. Keep adding stain as the slide cools, since the stain continues to evaporate.
6. Wash with water.
7. Holding the slide at an angle, carefully add the acid-alcohol solution one drop at a time to the smear until the red color stops coming off the smear. Immediately rinse with water to stop the decolorizing action.
8. Counterstain with Loeffler's methylene blue for 1 minute.
9. Rinse the slide **quickly** with water. Too long a rinse removes the blue dye from the cells.
10. Carefully blot the stained slide using a paper towel. Do *not* rub.

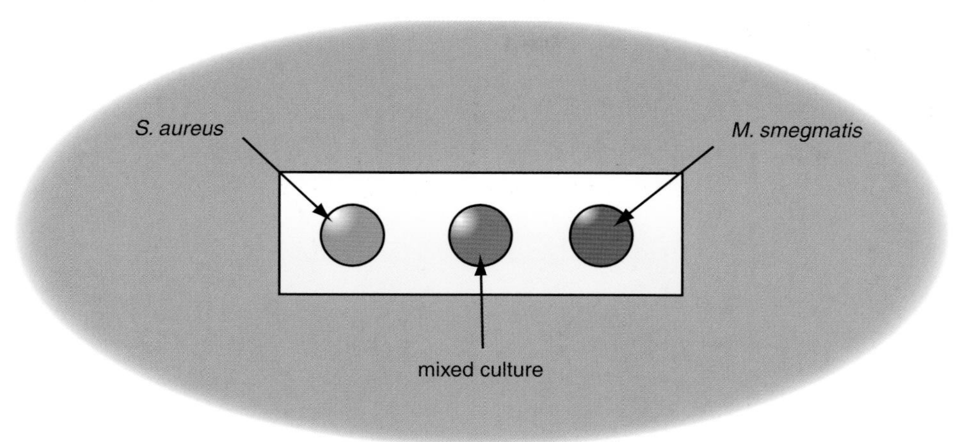

Figure 13-1

Preparation of acid-fast stain slide

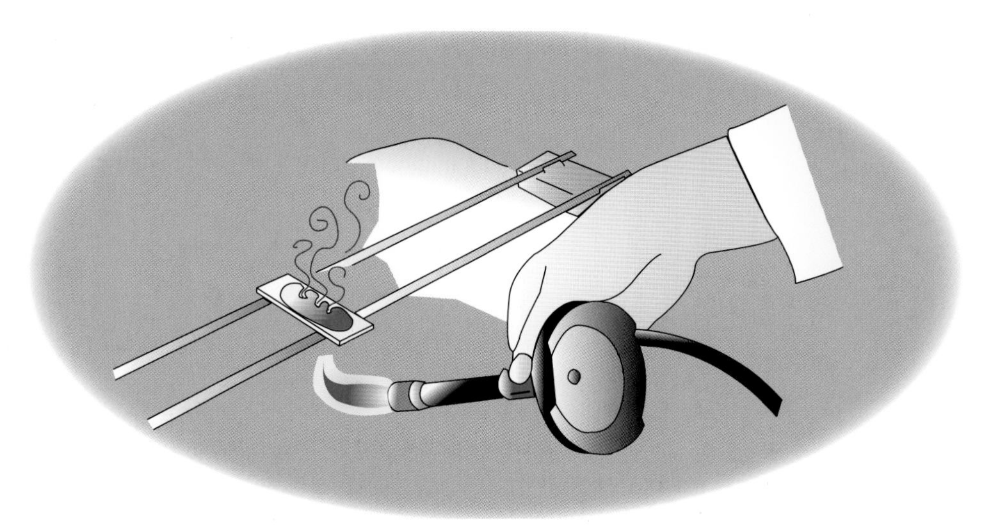

Figure 13-2

Steam but avoid boiling

Observations

Period 1

1. Examine with the oil immersion objective. Draw representative cells only from the mixed cell smear in the circle provided on the laboratory report form. Because acid-fast organisms tend to clump together, it may be necessary to examine many fields to find the cells. Use the pure culture smears for reference only.

14
Endospore Stain

Objectives

The student will be able to:

1. perform an endospore stain.
2. list the reagents used and the method of application for each step of the procedure.
3. describe the purpose and result of applying each reagent.
4. name two of the genera in which endospores are found.
5. name two diseases caused by endospore formers.
6. describe the morphology of endospore formers.

Species of the Gram-positive genera *Bacillus* and *Clostridium* and a few others produce an alternate cell type referred to as the **endospore**, often simply "the spore". Unlike the vegetative cell producing it, the endospore is highly resistant to a variety of physical and chemical agents, such as high temperatures, drying, radiation, and disinfectants. The endospore is *not* formed as a response to adverse conditions but is usually due to a change in the nutritional environment. Trace amounts of manganese as a nutrient stimulate endospore formation. The endospore is not a reproductive structure; it is a part of the life cycle of these spore-forming bacteria. The location, shape, and swelling of the cell (Figure 14-1) are useful characteristics in identifying the species of this group. The endospore-forming bacteria can be aerobic (*Bacillus*) or anaerobic (*Clostridium*). Some are pathogenic, with the best known being *Bacillus anthracis*, the cause of anthrax; *Clostridium tetani*, the cause of tetanus; and *Clostridium botulinum*, the cause of botulism.

The nature of the endospore requires vigorous treatment for staining. Once stained, the endospore resists decolorization and counterstaining. The **Schaeffer-Fulton** method uses *malachite green* as the primary dye, using heat to drive the dye into the endospore. After a decolorizing wash with *water*, *safranin* is applied as the counterstain. Vegetative cells accept the counterstain and appear red, while the endospores appear as small, green structures of various sizes and shapes.

Materials

1. 18- to 24-hour slant culture of *Staphylococcus aureus*
2. 12- to 18-hour slant culture of *Bacillus cereus* on manganese agar
3. 5% aqueous malachite green
4. 0.5% aqueous safranin
5. Slide holder or clothespin
6. Marking pen
7. Microscope slides
8. Staining rack
9. Paper towels

Procedure

Period 1

1. Prepare a mixed smear of *Staphylococcus aureus* and *Bacillus cereus* in the center of a slide. Place pure culture smears of each next to the mixed smear. Air dry and heat fix.
2. Cut a piece of paper toweling to cover the smear, yet not hang over the sides of the slide. Place the paper strip on the slide (Figure 14-2).
3. Saturate the paper and smear with malachite green, and allow it to stand for about 1 minute (Figure 14-3).
4. Heat the preparation to steaming by inverting your Bunsen burner and passing the flame over the stain or under the slide, moving back and forth the width of the stain (Figure 14-4).
5. Remove the burner when you see steam rising from the stain. When the steam stops rising, pass the flame over or under the stain again as necessary to keep the smear just at steaming. Steam for 5 minutes. Do not boil or allow the smear to dry. Add malachite green as needed to prevent loss of the stain by evaporation.
6. Allow the slide to cool to prevent breaking. Keep adding stain as the slide cools, since the stain solvent continues to evaporate.
7. Wash with water. Discard paper toweling in the trash, **not** in the sink.
8. Apply the 0.5% safranin counterstain, and allow it to react for 30–60 seconds.
9. Wash with water and blot dry using a paper towel. Do not rub.

Observations

Period 1

1. Examine with the oil immersion objective. Draw representative cells in the circle provided on the laboratory report form. Note

the position of the spore (green) within the red vegetative cell. Also label the different structures appropriately (i.e., endospore, vegetative cell).

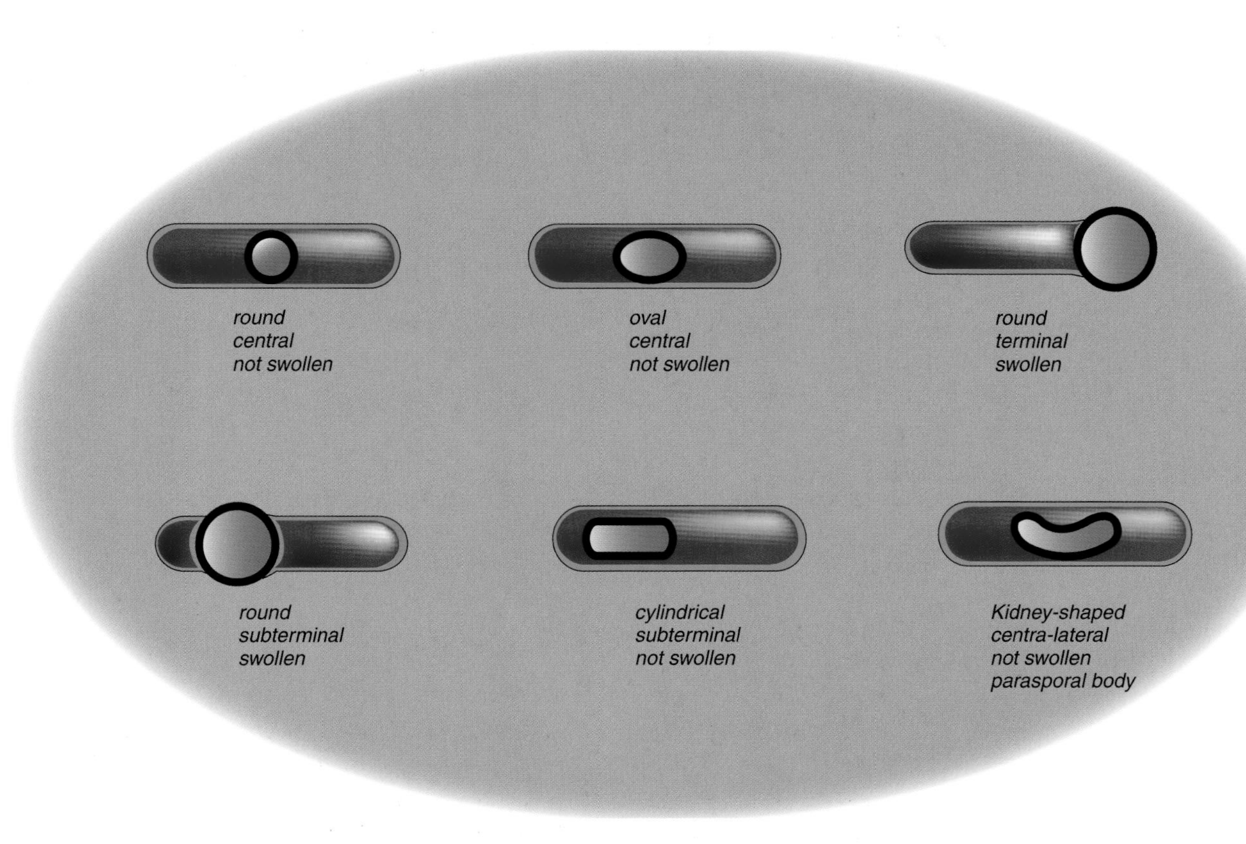

Figure 14-1

Endospores—shape, location, and swelling. Intermediate locations and shapes often occur.

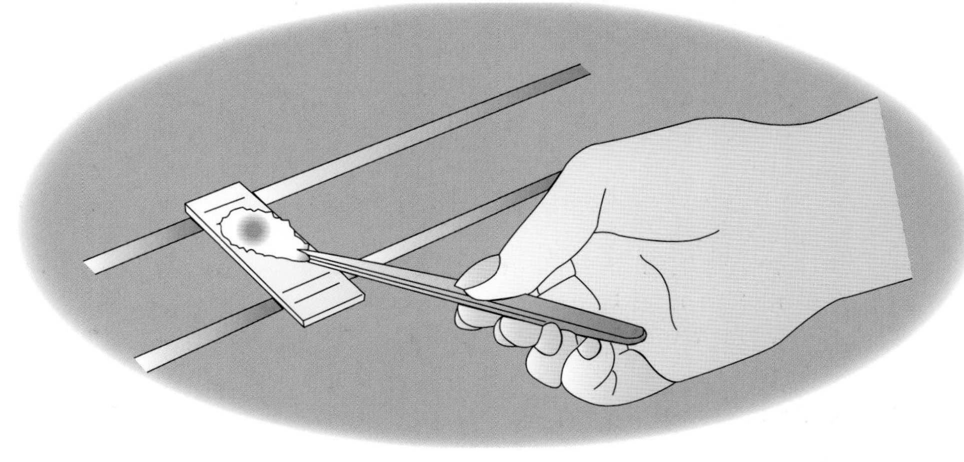

Figure 14-2

Apply paper towel

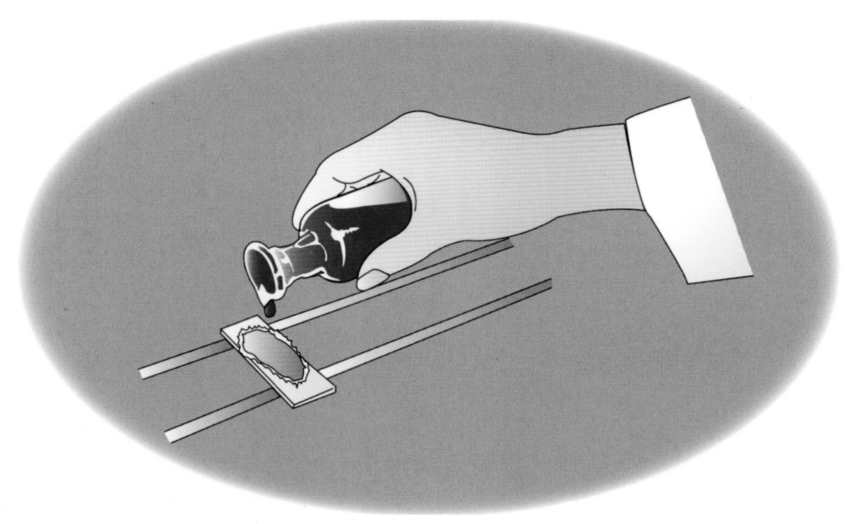

Figure 14-3

Add malachite green

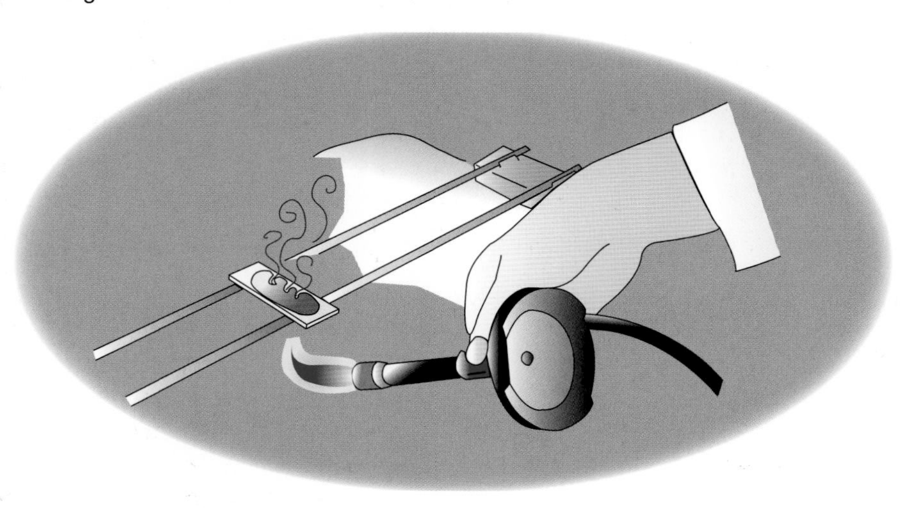

Figure 14-4

Steam, but avoid boiling

UNIT IV

Metabolic Activities

Bacteria are rather limited in morphological details and, except for some forms such as rods, cocci, and a few others, not very useful in distinguishing one from another. Instead, the most useful and important aspect of bacteria is what they do metabolically (i.e., the nature of their metabolic processes and products). In this section, a number of widely used metabolic tests are studied. The total examined here is a small fraction of those available, but these are some of the most universally used. These metabolic activities supply considerable information about the genetics, structure, and metabolism of the organisms. Although not widely used in identifying the higher microbes, except for the fungi, they still have some use even there. For the most part, these tests are used in classifying and identifying the bacteria and to some extent the fungi. Many of these tests have been combined in "multitest strips" of various kinds (see Exercise 34) for identification purposes. In order to understand more clearly what the test is and what it measures, they are presented separately here. If an unknown is to be studied, it might be useful to give it to the student and study it along with these tests. Coordinate this with Exercise 33 or 34.

Unit IV has been divided into four sections. The first, *Polymer Hydrolysis*, illustrates the ability of some bacteria to break down or hydrolyze large polymeric substances such as proteins, lipids, and starch, as well as DNA and blood cells (a special example of a protein). The nature of the cell wall in bacteria excludes these substances from moving into the cell. Microbes that can utilize these substances must form exoenzymes on the cell surface that may or may not be released into the environment. This is often visualized as a clearing of the polymer some distance from the colony of cells.

The second section deals with *Fermentation Reactions* (i.e., reactions that occur anaerobically—with limited oxygen availability) which result in partially oxidized end products from carbohydrates (sugars) or other organic or inorganic substances. Positive reactions generally indicate the organism has the ability to form the enzymes necessary to use these substances.

The third section, *Respiratory Reactions*, gives some information about the ability of the organism to hydrolyze hydrogen peroxide (nearly universal in aerobic organisms), form the enzyme oxidase, or use inorganic substances other than oxygen as terminal electron acceptors, thus respiring anaerobically.

The last section deals with some miscellaneous reactions.

It should be noted that most of the tests performed in this section can be done in special multitest units simultaneously. Please refer to Exercise 34 for a description of these. If multitest units are used in place of these exercises, the student should be familiar with the introductory material to these exercises, since they explain the microtest results as well.

15 Starch Hydrolysis

Objectives

The student will be able to:

1. give the general name of the enzyme resulting in the hydrolysis of starch.
2. write a general word equation illustrating a positive starch hydrolysis test.
3. name the chemical reagent used to test for the presence or absence of starch.
4. describe the appearance of a positive starch hydrolysis test.

Starch is a **polysaccharide**, a complex polymer of glucose found as a storage product in plants and many microbes. Microbes may possess one or more enzymes capable of hydrolyzing the starch polymer. These enzymes are collectively known as **amylases**.

Starch is an excellent source of energy and carbon for microbes but is too large to pass through the cell membrane. In order to utilize the energy in starch, an organism must excrete an extracellular amylase, an **exoenzyme** which may be attached to the cell surface or released into the environment, diffusing away from the cell. The exoenzyme cleaves the starch polymer outside the cell, and the smaller products can then be transported across the membrane. There are a number of kinds of amylases, depending on the site of activity in the starch molecule and the organism. Alpha-amylase cleaves starch to oligosaccharides, maltose, and glucose; beta-amylase cleaves starch to dextrins and maltose; and glucoamylase cleaves starch to glucose. These amylases are very important industrially and are harvested commercially from microbes for use in brewing, sizing of linen, and processing of paper, among others.

In this exercise, amylase activity is detected by the use of iodine to form a blue or brown complex with intact starch molecules. Where starch has been hydrolyzed, no color complex forms, appearing as a clear zone in a field of color. It is not possible with this technique to determine the specific type of amylase.

Materials

1. 18- to 24-hour cultures of *Bacillus subtilis* and *Escherichia coli*
2. 1 prepoured starch agar plate
3. Gram's iodine

Procedure

Period 1

1. Using a marking pen, divide the bottom of the plate into three equal segments.
2. Streak-inoculate one segment of the plate with *Bacillus subtilis* and another segment with *Escherichia coli*. Leave the third segment as an uninoculated control. Each streak should be about 1 cm long near the center of the segment (Figure 15-1).
3. Incubate at 37°C for 48 hours.

Observations

Period 2

1. Flood the surface of the plate with a thin layer of Gram's iodine and look for a color change of the medium. If the starch has not been hydrolyzed, the iodine reacts with the intact starch to give a brown, blue, or bluish-black color. When starch is hydrolyzed, the cleavage products no longer give the color reaction, and a clear zone of hydrolysis is observed.
2. Sketch the appearance of growth and the surrounding medium for each segment. Also complete the table provided.

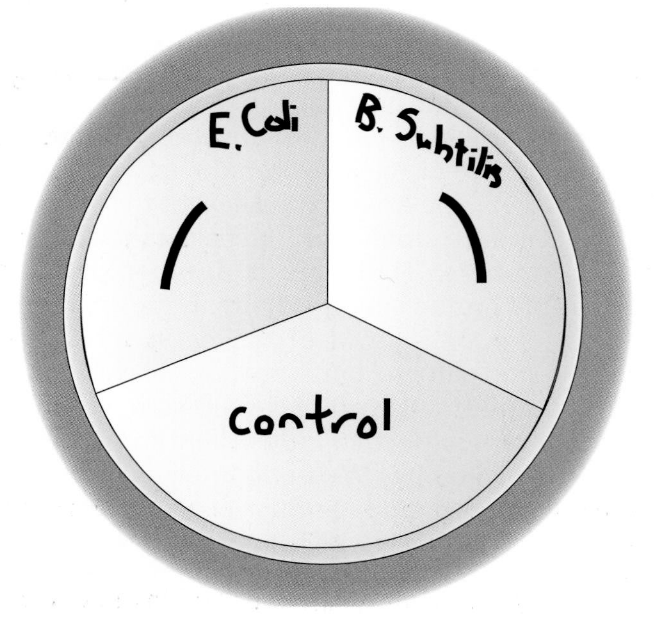

Figure 15-1

Inoculation of the medium

16
Lipid Hydrolysis

Objectives

The student will be able to:

1. give the general name of the enzyme that hydrolyzes fats and lipids.
2. write a general word equation illustrating a positive lipid hydrolysis test.
3. describe the two main ingredients of the medium used to detect lipid hydrolysis.
4. describe the appearance of a positive lipid hydrolysis test.

Lipids are large polymers, generally **hydrophobic** (repelling water), and usually containing an alcohol such as glycerol and one or more fatty acids in ester linkage or complex with phosphate, nitrogen, sulfur, and other carbon compounds. A fat, such as olive oil with three fatty acids esterified to glycerol, is a typical lipid as is the membrane lipid of cells (Figure 16-1).

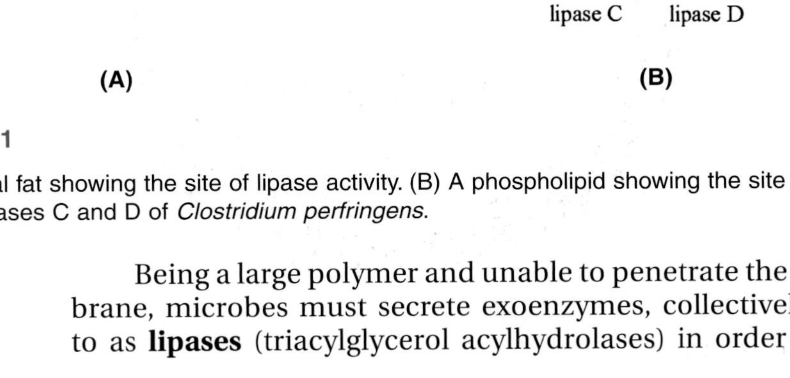

H_2C—O— fatty acid
HC—O— fatty acid
H_2C—O— fatty acid
glycerol
lipase

(A)

H_2C—O— fatty acid
HC—O— fatty acid
H_2C—O—P(=O)(OH)—O—X
lipase C
lipase D

(B)

Figure 16-1

(A) A typical fat showing the site of lipase activity. (B) A phospholipid showing the site of activity of phospholipases C and D of *Clostridium perfringens*.

Being a large polymer and unable to penetrate the cell membrane, microbes must secrete exoenzymes, collectively referred to as **lipases** (triacylglycerol acylhydrolases) in order to reduce the molecule to units able to enter the cell. Once inside the cell, fatty acids can be used for energy or as building blocks for biosynthesis. Many lipases are not very specific and attack fats with side chains of various lengths. Some, such as the phospholipases of *Clostridium perfringens*, hydrolyze the phospholipid found in the blood cell leading to lysis of the cell (hemolysis) and producing some of the symptoms of gas gangrene. The hydrolytic products can then be transported across the cell membrane for use in metabolism.

In this exercise, the hydrolytic products of lipase activity are visualized by uptake of a blue dye, resulting in an intense blue color.

Materials

1. 18- to 24-hour cultures of *Staphylococcus aureus* and *Proteus mirabilis*
2. 1 prepoured Spirit blue-lipid agar plate

Procedure

Period 1

1. Using a marking pen, divide the bottom of the plate into three equal segments.
2. Streak-inoculate one segment of the plate with *Staphylococcus aureus* and another segment with *Proteus mirabilis*. Leave the third segment as an uninoculated control. Each streak should be about 1 cm long near the center of the segment (Figure 15-1).
3. Incubate at 37°C for 48–72 hours.

Observations

Period 2

1. Observe for the presence of a clear, somewhat darker blue color in or around the growth. Organisms that utilize lipid by producing lipase cause an intense blue precipitate to form in or under the growth. This is due to the release of fatty acids and the uptake of Spirit blue. If no lipase has been produced, no darker color change develops. You may need to scrape some growth away with a loop to look under the streak.
2. Record your results in the table provided.

17 Protein Hydrolysis (Gelatin and Casein)

Objectives

The student will be able to:

1. give the general names of the enzymes that hydrolyze gelatin and casein.
2. write a general word equation illustrating positive gelatin and casein hydrolysis tests.
3. define the terms *proteolysis* and *peptonization*.
4. name the chemical reagent used to test for the presence or absence of gelatin or casein using the plate overlay method.
5. describe the appearance of positive tests for gelatin and casein hydrolysis using the plate overlay method.

Many microorganisms produce exoenzymes capable of hydrolyzing large proteins (polymers of amino acids) into soluble end products of various sizes called peptones, proteoses, or peptides, and the constituent amino acids. The smaller fragments can be transported across the membrane into the cell for use in building cell material or for energy. The enzymes are collectively referred to as **proteases**. Some are named after the protein **substrate** such as gelatinase, collagenase, and so forth. The enzymes are highly specific for certain proteins or amino acid bonds within the molecule. The degradation process is variously called **proteolysis** or **peptonization**. It is not the intention here to explore the details of these enzymes but to present a general method used to detect enzymes capable of degrading proteins. Two protein substrates are used here: gelatin, a typical animal protein, and casein, a milk protein.

Gelatin is a protein derived from collagen found in hide and connective tissue in animals. A 12% gelatin solution is a solid (gel) at room temperature (20°C) but melts at 28°–30°C to form a solution. Degradation of gelatin leads to a loss of the gelling property at room temperature and was the basis of a crude test used in early microbiology. A nutrient gelatin medium was inoculated and incubated. After a few days the tube was refrigerated. A failure to gel indicated that the organism produced a protease also called a **gelatinase**. This is not a very sensitive test (rarely used today) and was replaced by a very sensitive overlay method (Fraser) used in this exercise. An even more sensitive and quantitative method for collagen degradation (**collagenase** enzyme) involves the use of a dye coupled to the collagen with subsequent measurement of dye release using the spectrophotometer.

Casein is a protein restricted to animal milk. Historically and industrially, the ability of an organism to hydrolyze casein has proven very useful. One of the oldest microbiological tests involved the used of a tube of milk with the addition of the dye litmus (litmus milk). An organism growing in the milk capable of digesting the casein causes clearing of the milk, a process referred to as **peptonization**. Again, this is not a very sensitive method (presented in Exercise 18) and can be replaced by the overlay method mentioned above.

The overlay method of Fraser described here relies on the ability of a protein to form a colloid (white precipitate) or to be precipitated by certain chemicals, such as acidified mercuric chloride or trichloracetic acid. Acid alone will often cause precipitation. A protein is added to a small amount (5 ml or so) of a nutrient agar medium and overlaid on the surface of a prepoured nutrient medium. The amount of protein must be sufficient to be seen as a cloudy layer (casein) or sufficient to react with the precipitating reagent (gelatin). If casein is hydrolyzed, it will appear as a clear zone around the growth. The chemical precipitant may enhance the cloudy appearance of intact casein. If gelatin is hydrolyzed, it is seen by adding the chemical reagent causing unhydrolyzed gelatin to precipitate. Your instructor may elect to use only one of these substrates.

Materials

1. 18- to 24-hour broth cultures of *Serratia marcescens*, *Bacillus cereus*, and *Escherichia coli*
2. 2 prepoured nutrient agar plates
3. 1 bottle skim milk-nutrient agar for 15 students, sterile
4. 1 bottle nutrient-gelatin agar for 15 students, sterile
5. Acidified mercuric chloride solution or other protein-precipitating reagent

Procedure

Period 1

1. Overlay one nutrient agar plate with 5 ml of the skim milk-nutrient agar and allow it to solidify. Label it "casein".
2. Overlay the second nutrient agar plate with 5 ml of nutrient-gelatin agar and allow it to solidify. Allow the plate to set about 30 minutes so that the agar will set and the surface will become relatively dry. Label it "gelatin".
3. Using a marking pen, divide the bottom of each plate into three equal segments (Figure 15-1).
4. Streak-inoculate one segment of the "casein" plate with *Bacillus cereus* and another segment with *Escherichia coli.* Leave the third segment as an uninoculated control. Each streak should be about 1 cm long near the center of the segment.
5. Streak-inoculate one segment of the "gelatin" plate with *Serratia marcescens* and another segment with *Escherichia coli.* Leave the third segment as an uninoculated control. Each streak should be 1 cm long near the center of the segment.
6. Incubate the "casein" plate at 32°C for 48–72 hours.
7. Incubate the "gelatin" plate at 37°C for 48–72 hours.

Observations

Period 2

Gelatin

1. Flood the "gelatin" plate with the solution of acidified mercuric chloride and let the plate stand for 5–10 minutes.
2. Wherever **unhydrolyzed** gelatin remains, this protein precipitates, forming a cloudy white turbidity. If the gelatin has been hydrolyzed by the action of the gelatinase, the precipitate does **not** appear since the hydrolyzed gelatin products cannot precipitate.
3. Sketch the appearance of growth and the surrounding medium for each segment on your report form.

Casein

1. After incubation, enzyme activity is read directly on the "casein" plate. Areas in which the casein has **not** been attacked will remain **opaque** due to the colloidal nature of milk protein. If the enzymes have been produced, a clear zone appears around the area in which the organism has grown. You can usually see the clear zones best against a dark or black background. Clearing of the milk is a positive test.
2. The plate may be flooded with the acidified mercuric chloride solution to accentuate the reaction, if desired, but is usually not necessary.
3. Record your results in the report form table provided.

18 Litmus Milk Reactions

Objectives

The student will be able to:

1. list the two main ingredients of litmus milk.
2. write a general word equation illustrating the reactions resulting from bacterial action on litmus milk:
   a) acid production
   b) gas production
   c) curd formation (two types)
   d) peptonization
   e) dye reduction

One of the earliest diagnostic media used to characterize the metabolic activity of bacteria was litmus milk. Milk makes an excellent culture medium for most bacteria and, with the addition of the dye litmus, can be used to make many biochemical observations. Although it is difficult to read and often is highly variable, even with the same culture, it is still widely used in the diagnostic laboratory, despite the development of newer and more sensitive methods. Litmus milk medium consists of 10% powdered skim milk and the dye litmus. When added to rehydrated skim milk, litmus turns the milk suspension from white to lavender and serves as both a pH indicator and as a reducible dye molecule.

Reduction. Dyes and pH indicators, such as litmus, appear colored as a result of their chemical bond structure as an oxidized molecule. Most of them are able to accept electrons (some easily, some with great difficulty), and in doing so many become colorless. Certain bacteria capable of growing anaerobically often are able to transfer metabolic electrons to the dye molecule in the absence of oxygen, thus reducing the dye in some cases to a colorless form. Not all bacteria are able to do this nor are all dyes so reduced. The ability to do this depends on the presence of enzymes of the right electromotive potential and the potential of the dye (i.e., ease of reducibility). Litmus is one of the easiest to reduce by many bacteria. If an organism is capable of reducing litmus, the lavender-colored milk turns to its normal white-milk color again. *Reduction* characteristically begins at the bottom of the tube and progresses in an upward direction. When litmus is in the reduced state, it can no longer function as a pH indicator, since there is no color to observe.

Acid. If lactose (the main carbohydrate in milk) is fermented, then the litmus acting as a pH indicator turns from lavender to a red or pink color in an acid environment. If the dye is reduced in the bottom of the tube, an acid-base color may be seen only at the top or as a ring around the surface. If the milk is still fluid, shaking the tube vigorously re-oxidizes the litmus, restoring the acid or basic color.

Acid Curd. If enough acid is produced, the pH may be lowered below the isoelectric point of casein and the milk forms a very hard curd. This is called an *acid curd* and may or may not be accompanied by observable gas production. As the acidity increases, the curd becomes so solid that there is a squeezing out of a clear liquid on top called *whey. Gas production* occurs only if an organism is capable of fermenting lactose to acid and gas. It is usually detected by the presence of bubbles or a splitting in the acid curd, although it often cannot be observed by this method. If an organism produces so much gas that the curd is blown into shreds, it is called *stormy fermentation* of milk.

Rennet Curd. Some organisms produce a rennin-like enzyme which clots casein gently, resulting in the formation of a *rennet curd.* The rennet curd, unlike the acid curd, is soft and usually observed only at a neutral pH.

Peptonization. Some microorganisms possess proteolytic enzymes capable of hydrolyzing the insoluble casein. This process, called *peptonization,* results in the release of large amounts of peptides and amino acids (also see Exercise 17). Continued incubation results in a clearing of the milk. The transparent liquid supernatant often turns brown. An organism may release ammonia, causing a purplish-blue color. A tube may often show acid production early in incubation which reverts to a basic reaction after peptonization, because the ammonia neutralizes the acid. Shaking peptonized tubes to re-oxidize the litmus is often the only way to determine the pH.

Materials

1. 18- to 24-hour cultures of *Escherichia coli*, *Pseudomonas aeruginosa*, *Acinetobacter calcoaceticus*, and *Enterococcus faecalis*
2. 5 tubes litmus milk, sterile

Procedure

Period 1

1. Inoculate each tube of litmus milk with one of the cultures.
2. Keep one tube as an uninoculated control.
3. Incubate all tubes at 37°C. Make your observations at 24 and 48 hours and again at 7 days.

Observations

Periods 2–4

1. At each incubation period, compare each of the inoculated tubes with the control tube. WARNING! Do NOT shake the tubes until the **last** observation day.
2. Make observations as follows:
   a. **Acid.** Look for a change in color. Red = acid; lavender = no change; deep purple = alkaline. Note that the color may be restricted to the surface layer only.
   b. **Reduction.** Look for a decolorization of the litmus at the bottom of the tube. Autoclaving the medium sometimes decolorizes the dye, but oxygen restores the color, with the possible exception of the very bottom of the tube. Comparison of the tubes over time is the only way to distinguish bacterial reduction from autoclave reduction. Be sure to record the level of the decolorization in the tube each time you make observations.
   c. **Acid Curd.** Tip the tube gently to the side. An acid curd will appear as a solid and the medium will not pour (except for whey on the top). Some bacteria form such a solid curd that the tube can be inverted without spilling. **Do NOT** do this, however. Record the presence of whey, if seen.
   d. **Rennet Curd.** This can be observed only in the absence of acid production. Tip the tube to the side and watch the fluid level. Rennet curds flow very gently, like a heavy syrup. Note that this is rarely seen.
   e. **Peptonization.** This appears as a gradual disappearance of the opaque casein suspension. It may begin at the top and work down the tube with time. Or, it may progress down the side of the tube, especially if there is a curd formed. Sometimes it may turn brown, but not always.
   f. At your **last** observation, if there is no curd, gently shake the tube to restore any pH indicator color. Do this only if the tube is entirely colorless.
3. Record your observations by code in the table provided.

19 Nuclease Activity

Objectives

The student will be able to:

1. perform a test for nuclease activity.
2. define nuclease.
3. explain the significance of the thermostable nuclease of *Staphylococcus aureus*.
4. describe, in a general way, the official Food and Drug Administration method.

Bacteria that form **exonucleases** (extracellular nucleases) can utilize the carbon, nitrogen, and end products of large polymers of RNA or DNA from the environment. In some bacteria, *Staphylococcus aureus* in particular, the nuclease is **thermostable** and withstands boiling temperatures. This property is useful in identifying *S. aureus* **food poisoning** organisms either in culture or in foods when heating has destroyed the organism but not the nuclease. The general procedure for detecting DNA or RNA hydrolysis is similar to other large polymers. The nucleic acid is incorporated into an agar medium, an organism applied, and incubated. The plate is then developed with 4N HCl which precipitates intact nucleic acid with hydrolysis appearing as a clear zone. A toluidine blue medium is also available for DNA in which hydrolysis appears as a pink zone in a blue unhydrolyzed background.

The Food and Drug Administration procedure for food studies prescribes a slide technique for thermonuclease (*FDA Bacteriological Analytical Manual*, 7th edition, 1992, Association of Official Analytical Chemists, Washington, DC). In this method, a thin film of toluidine blue-DNA agar is placed on a glass slide. Ten to twelve small 2 mm diameter wells are cut in the agar and several loopfuls of broth cultures or centrifuged food slurries are placed in each well. The slides are incubated for 4 hours and read. A pink to red zone around the well shows DNA hydrolysis.

This exercise is intended to illustrate the activity of nucleases.

Materials

1. 18- to 24-hour-old broth cultures of *Staphylococcus aureus*, *Staphylococcus epidermidis*, and *Serratia marcescens*
2. 1 prepoured nutrient agar plate
3. 1 tube toluidine blue-DNA (TB-DNA) agar for overlay, sterile

Procedure

Period 1

1. With a marking pen, divide the bottom of the nutrient agar plate into three segments. Label with the organism names.
2. Streak-inoculate (about 1 cm in length) the center of each segment with the corresponding organism.
3. Incubate the plate for 24–48 hours at 35°–37°C.

Period 2

4. Melt 5–10 ml of TB-DNA agar (the overlay tube), cool to 50°C, and pour over the growth on the plate. Allow it to solidify.
5. Incubate the overlay plate at 35°–37°C for 3–4 hours.

Observations

Period 2

1. After incubation, organisms showing DNA hydrolyzing activity will have a pink to red zone around the colony.
2. Record your results on the report form.

20 Hemolysis

Objectives

The student will be able to:

1. name the three types of hemolysis produced by microorganisms.
2. describe the appearance of the three different types of hemolysis.
3. list the two main ingredients of a blood agar plate.

Whole blood added to a nutrient agar makes a very rich medium which supports the growth of many of the more nutritionally fastidious microorganisms. Many such organisms produce exoenzymes, generally called **hemolysins**, that have a destructive effect on red blood cells and the hemoglobin inside. **Beta** (β) **hemolysins** completely destroy the blood cell and decolorize the hemoglobin, resulting in a clear zone around the colony. Some organisms produce **alpha** (α) **hemolysins** which only partially destroy the hemoglobin, resulting in a greenish, sometimes cloudy area around the colony. Organisms that do not produce hemolysins, and subsequently have no effect on the red blood cells, are sometimes said to demonstrate gamma hemolysis. More accurately, these are **non-hemolytic**. The production of these hemolysins is very useful in identifying pathogens, especially among the streptococci.

Materials

1. 18- to 24-hour cultures of *Enterobacter aerogenes*, *Staphylococcus aureus*, and *Enterococcus faecalis*
2. 5% sheep blood agar plate, prepoured

Procedure

Period 1

1. Using a marking pen, divide the bottom of the blood agar plate into four equal segments (Figure 20-1).

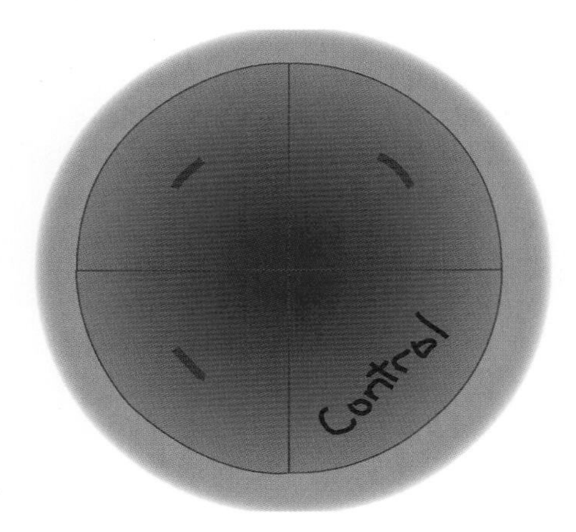

Figure 20-1

Inoculation of the medium

2. Streak-inoculate one segment of the plate with *Enterobacter aerogenes*. Repeat the procedure with each of the remaining organisms. Leave the fourth segment as an uninoculated control. (Each streak should be about 1 cm long and located near the center of the segment.)
3. Incubate at 37°C for 24 hours maximum. Note: Too long an incubation at this temperature may destroy the blood cells and negate the observations. If plates cannot be read at 24 hours, place them in the refrigerator for later reading.

Observations

Period 2

1. Observe for the different types of hemolysis (Figure 20-2) using a lighted Quebec colony counter.
2. Sketch the appearance of growth and the surrounding medium in each segment. Also complete the table provided.

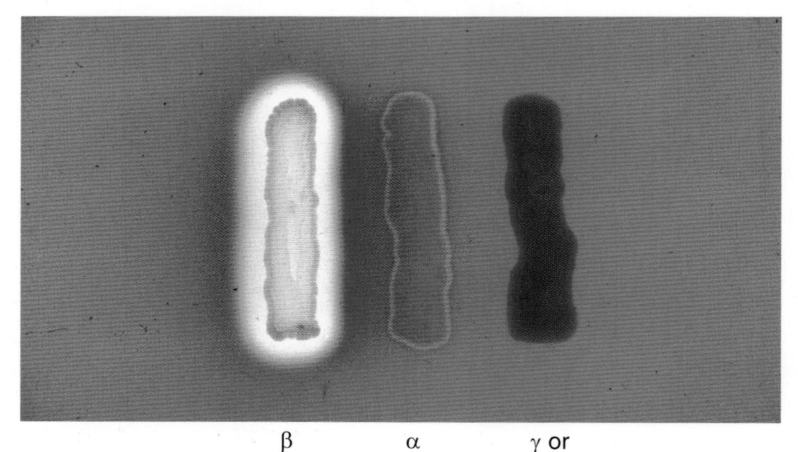

Figure 20-2

Examples of different types of hemolysis

21 Sugar Fermentations

Objectives

The student will be able to:

1. name the general group of enzymes involved in the fermentation of carbohydrates.
2. define the term fermentation.
3. write a general word equation illustrating the major end products of carbohydrate fermentation.
4. list the four ingredients and components of a fermentation tube and describe the purpose of each.
5. describe all possible reactions that can be obtained from microorganisms growing in fermentation broths.

Fermentation is the term ordinarily applied to the anaerobic breakdown of carbohydrates or sugars. The purpose of the fermentation process is to make energy available for utilization by the microorganism, since carbohydrates are rich in stored energy.

Whether or not a given carbohydrate is fermented depends upon the **transport proteins** (permeases) and the **endoenzymes** (or carbohydrases) possessed by the organism, and these are very specific. The end products of fermentation often vary from one organism to another depending on the pathways involved. However, the end products are usually acids of various types *or* acids *and* gas. Large carbohydrates such as starch must first be hydrolyzed to units small enough to be carried by transport proteins into the cell.

Ability to ferment can be determined by inoculating the organism into a fermentation tube containing a nutrient broth to su pport the growth of the organism, a single chemically-defined carbohydrate, a pH indicator, and an inverted Durham tube to collect gas.

pH indicators allow you to determine whether or not an **acid** has been produced as an end product of metabolism. Phenol red (PR) and bromcresol purple (BCP) are the two pH indicators most

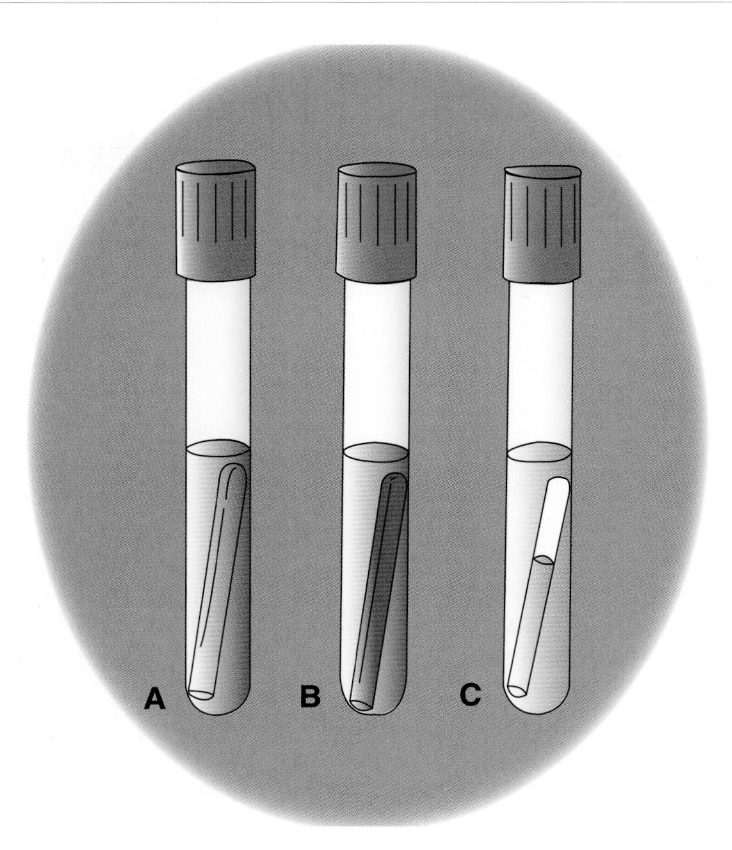

Figure 21-1

Broth culture tubes with Durham tubes to measure gas production: (A) before inoculation; (B) growth but no gas production; (C) growth resulting in gas production.

frequently used in microbiological work. Phenol red appears red in a solution with a pH above 6.9 and is yellow at an acid pH of less than 6.8. Bromcresol purple is purple at a pH of 6.8 and is yellow at an acid pH of 5.2 or lower.

A Durham tube is a small inverted vial placed inside the culture tube. The purpose of the Durham tube is to trap **gas**, if it has been produced as an end product of metabolism.

If the organism is not capable of attacking the specific carbohydrate, growth will occur due to the utilization of the other nutrients present. No change in pH or a change to a more basic pH will be seen in these cases.

Some organisms can ferment amino acids in the broth to carbohydrate-like molecules called alpha-keto acids and also results in the formation of ammonia (NH_3) which, in the absence of carbohydrate, causes the medium to become quite basic or alkaline. The bromcresol purple indicator will change to a deeper purple color, while the phenol red indicator turns a deep magenta color.

Materials

Note: *Label each tube as you take it, since they cannot be distinguished from each other.*

1. 18- to 24-hour cultures of *Escherichia coli*, *Staphylococcus aureus*, *Bacillus subtilis*, and *Alcaligenes faecalis*
2. 5 tubes phenol red-glucose broth with gas insert, sterile
3. 5 tubes phenol red-sucrose broth with gas insert, sterile
4. 5 tubes phenol red-lactose broth with gas insert, sterile

Procedure

Period 1

1. Inoculate a series of three different carbohydrate broths with *Escherichia coli.*
2. Repeat this procedure with each of the remaining cultures.
3. Keep one tube of each carbohydrate broth as uninoculated controls.
4. Incubate all tubes at 37°C.

Observations

Period 2–5

1. Make your observations at 24, 48, and 72 hours and again at 7 days.
2. At each observation time, compare each of the inoculated tubes with the control (uninoculated) tube of the same medium to determine whether growth (turbidity) occurred and whether acid or acid and gas were produced.
   If at any time during the incubation period a fermentation tube is found to contain both acid and gas, it is not necessary to continue incubation. The tube may be discarded after recording the results.
3. Record your results in the table provided using the codes given on the report form.

22 Hydrogen Sulfide Production

Objectives

The student will be able to:

1. name one component of a substrate that must be present for a microorganism to produce hydrogen sulfide.
2. write a general word equation illustrating a positive hydrogen sulfide test.
3. list four types of media designed to show hydrogen sulfide production.
4. describe the appearance of a positive hydrogen sulfide test using Kligler iron agar.
5. describe all possible reactions that can be observed from the fermentation of carbohydrates in Kligler iron agar.

Hydrogen sulfide (H_2S) is a gas produced by certain microorganisms through **dissimilation** of organic sulfur-containing amino acids (cystine, cysteine, and methionine) under anaerobic conditions or through **reduction** of inorganic sulfur compounds (thiosulfate, sulfite, and sulfate). Such organisms possess enzymes that reduce the sulfur atom of inorganic sulfur-containing compounds or remove the sulfide group from sulfur-containing amino acids. Ferrous iron salts (e.g., $FeSO_4$) react readily with hydrogen sulfide to form black iron sulfide (FeS). In nature, highly anaerobic muds in swamps, lakes, and streams are often colored a deep black because of this reaction.

In the laboratory, hydrogen sulfide production can be detected by incorporating a ferrous salt into the medium. Various types of media have been designed for the purpose of detecting hydrogen sulfide production. Among these are peptone iron agar (PIA), Kligler iron agar (KIA) [Figure 22-1], and triple sugar iron agar (TSI). Lead salts (e.g., lead acetate) can also be used but are somewhat toxic. Lead acetate-impregnated paper strips suspended over a broth culture can detect the gas when it turns black.

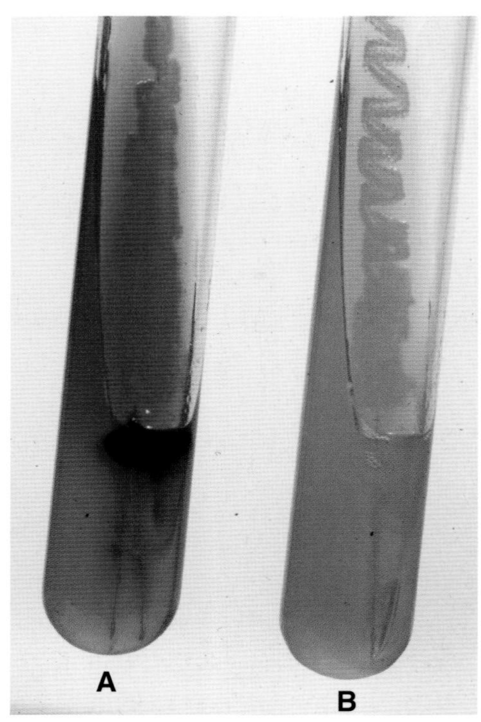

Figure 22-1

KIA slants: (A) positive hydrogen sulfide and (B) negative hydrogen sulfide tests

Both peptone iron agar and lead acetate agar are employed to demonstrate only hydrogen sulfide production. Since lead acetate has been shown to be somewhat less sensitive than peptone iron agar, it is not frequently used to detect hydrogen sulfide production. Both Kligler iron agar and triple sugar iron agar are considered multipurpose media and are used to demonstrate the fermentation of certain carbohydrates as well as the production of hydrogen sulfide. These media were designed for use in identifying pathogens. Except for hydrogen sulfide production, sugar fermentation methods are not reliable at incubation temperatures other than 37°C.

Materials

1. 18- to 24-hour slant cultures of *Proteus vulgaris* and *Escherichia coli*
2. 3 Kligler iron agar (KIA) slants, sterile

Procedure

Period 1

1. Using an inoculating needle and the culture of *Proteus vulgaris,* penetrate the butt of a Kligler iron agar slant about the center of the slant nearly, but not all the way, to the bottom of the tube (Figure 22-2A). Carefully withdraw the needle along

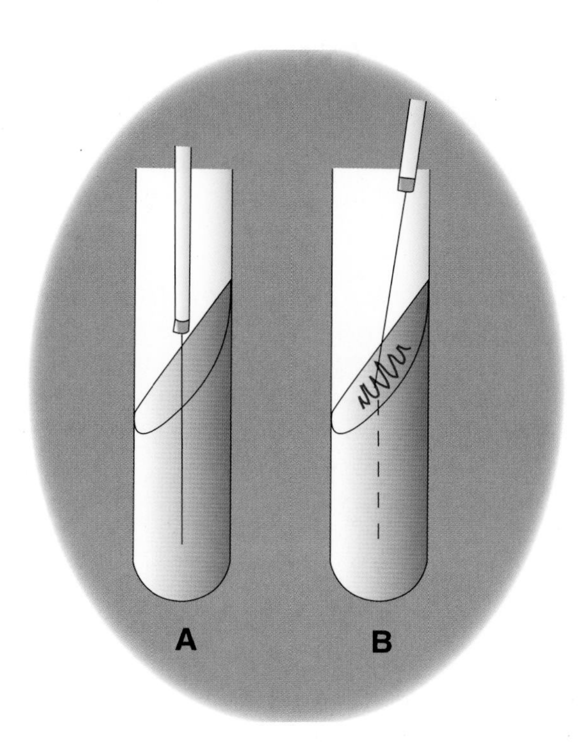

Figure 22-2

Stabbing (A) and streaking (B) a KIA slant

the original stab and then streak the culture across the top of the slant (Figure 22-2B).

2. Repeat the procedure with the culture of *Escherichia coli*.
3. Keep one tube as an uninoculated control.
4. Incubate all tubes at 37°C.

Observations

Periods 2–5

1. Make your observations at 24 and 48 hours and again at 4 and 7 days.
2. After the 24-hour incubation period, make observations on carbohydrate fermentation reactions by comparing each inoculated tube with the control. If both the slant and the butt remain red (due to phenol red indicator), then no change has occurred. In other words, no carbohydrates have been fermented. If the slant is red and the butt is yellow, this indicates that glucose has been fermented to acid. If gas bubbles or splitting appears in the butt, then the glucose has been fermented to acid and gas. If both the slant and butt are yellow, then both glucose and lactose have been fermented. The appearance of gas in the butt indicates that

these carbohydrates have been fermented to acid and gas. Record your results in the table provided.

(*Note:* Fermentation reactions are valid only at 37°C and at 24 hours. After 24 hours, the colors are no longer useful since they may have reverted to alkaline colors. Also, a large amount of black precipitate can mask or obscure the yellow or acid condition in the butt.)

3. After each incubation period, observe for the appearance of a black precipitate along the line of the stab. Such a black precipitate is evidence of hydrogen sulfide production. No black in the medium means that no hydrogen sulfide was produced. Record your results in the table provided.

23
IMViC Reactions

Objectives

The student will be able to:

1. distinguish *Escherichia coli* from *Enterobacter aerogenes* by IMViC tests.
2. explain the biochemical basis for each IMViC test.
3. write down the IMViC reactions for *Escherichia coli* and *Enterobacter aerogenes.*
4. name the chemical reagent(s) used and the appearance of a positive test result for each IMViC test.

The IMViC reactions are four related biochemical tests. The acronym **IMViC** is a mnemonic to aid in remembering the order of the four tests and their reactions. They are **Indole** production, the **Methyl red** test, the **Voges-Proskauer** test, and **Citrat** utilization. The small letter "i" is added as an aid in pronouncing the acronym.

The IMViC tests are designed to determine specific physiological properties of microorganisms. They are especially useful in the differentiation of a special subset of the Gram-negative enteric bacilli known as the coliforms, widely used as indicators of fecal contamination. Coliforms include the *Escherichia coli* and the *Enterobacter-Klebsiella* groups. Because members of the coliforms have very similar physiological reactions and often cannot be adequately differentiated, the IMViC tests are a considerable aid to identification of organisms within the coliform group. Some of the tests are also useful in identifying other members of the family *Enterobacteriaceae.*

Indole is a breakdown product of the amino acid tryptophan, which is found plentifully in tryptone or trypticase. It is important that the test be read after the 24-hour incubation, since continued incubation may lead to a loss of indole by further metabolism. Should this occur, a misleading negative test may result. Typical strains of *E. coli* produce indole, while the *Enterobacter-Klebsiella* group does not.

The **methyl red** and **Voges-Proskauer** tests are physiologically related and both use the same medium, MR-VP broth. This medium contains peptone, glucose, and a phosphate buffer. The methyl red test depends upon the ability of an organism to produce acid from glucose in sufficient quantity to cause the methyl red indicator to change to its acid color red and hold this low pH of 4.4 for 5 days. Typical strains of *E. coli* are mixed acid fermenters, which produce a variety of acid end products which change the methyl red indicator to red. The *Enterobacter-Klebsiella* organisms ferment the same amount of glucose but convert it over several days into neutral products such as acetoin (acetyl methyl carbinol) and 2,3 butanediol. Consequently, the pH, which may initially reach pH 4.4, very quickly rises again, and the methyl red no longer displays its acid red color.

The **Voges-Proskauer** reaction is the result of a red color complex formed by the oxidation of 2,3 butanediol and acetoin in alkali and the presence of creatine and alpha-naphthol. *E. coli* does not produce these end products and is Voges-Proskauer–negative. Since the *Enterobacter-Klebsiella* group produces 2,3 butanediol and acetoin, it produces a positive (red) Voges-Proskauer test.

The last IMViC test determines whether an organism can grow with **citrate** as a sole source of carbon. The test is designed to determine whether or not the bacterium has a cell membrane transport protein to carry citrate into the cell. If the organism has the particular protein, then it grows. If the organism lacks the protein, then it cannot grow. *E. coli* does not utilize citrate, whereas members of the *Enterobacter-Klebsiella* group do. Either a liquid (Koser's citrate broth) or a solid medium (Simmon's citrate agar) can be used.

It should be noted that organisms exist that exhibit all combinations of IMViC results. Those not typical *E. coli* (+ + – –) or typical *E. aerogenes* (– – + +) are referred to as **intermediates**.

Materials

1. 18- to 24-hour cultures of *Escherichia coli* and *Enterobacter aerogenes*
2. 3 tubes 1% tryptone broth, sterile
3. 6 tubes MR-VP broth, sterile
4. 3 Simmons citrate agar slants, sterile
5. Kovac's indole reagent
6. Methyl red indicator
7. 5% alpha-naphthol (Barritt's solution A)
8. 40% KOH - creatine reagent (Barritt's solution B)

Procedure (5 days)

Period 1

A. Indole Test

1. Inoculate one tube of tryptone broth with a small amount of *E. coli.*
2. Inoculate a second tube with the culture of *E. aerogenes.*
3. Keep one tube as an uninoculated control.
4. Incubate all tubes at 35°C for 24 hours only.

B. Methyl Red Test

1. Inoculate one tube of MR-VP broth with *E. coli.*
2. Repeat the procedure with the culture of *E. aerogenes.*
3. Keep one tube as an uninoculated control.
4. Incubate all tubes at 35°C for 5 days.

C. Voges-Proskauer Test

1. Inoculate one tube of MR-VP broth with *E. coli.*
2. Repeat the procedure with the culture of *E. aerogenes.*
3. Keep one tube as an uninoculated control.
4. Incubate all tubes at 35°C for 48 hours.

D. Citrate Utilization

1. Using an inoculating needle and the culture of *E. coli,* penetrate the butt of a Simmon's citrate agar slant about halfway through the agar. Carefully withdraw the needle along the original stab, and then streak the culture across the top of the slant. A small amount of inoculum is essential. (See Figure 22-2 for method.)
2. Repeat the procedure with the culture of *E. aerogenes.*
3. Keep one tube as an uninoculated control tube.
4. Incubate all tubes at 35°C for 48 hours.

Observations

A. Indole Test

Period 2

1. After incubation for 24 hours only, compare each of the inoculated tubes with the control tube.
2. Examine each tube for growth. Add 0.2–0.3 ml (approximately 5 drops) of Kovac's indole reagent to each tube.

Shake gently and let stand. Within 10 minutes, the amyl alcohol of the reagent will separate and turn a bright red color if indole is present. If indole is absent, the amyl alcohol layer will remain as its original color.

3. Record your results in the table provided.

B. Methyl Red Test

Period 4

1. After incubation for 5 days, compare each of the inoculated tubes with the control tube.
2. Examine each tube for growth. Shake lightly and add 4–5 drops of methyl red indicator. A red color indicates acid below pH 4.4 and is a positive test. A yellow color is a negative test.
3. Record your results in the table provided.

C. Voges-Proskauer Test

Period 3

1. After incubation for 48 hours, compare each of the inoculated tubes with the control tube.
2. Add 1 ml of Barritt's solution A followed by 0.5 ml of Barritt's solution B. A red positive color will usually develop promptly, but may be delayed up to 2 hours. No change in color constitutes a negative test.
3. Record your results in the table provided.

D. Citrate Test

Period 3

1. After incubation for 48 hours, compare each of the inoculated tubes with the control tube.
2. Organisms able to transport citrate into the cell grow in the medium. In the Simmon's medium, growth is observed on the slant and is usually accompanied by a change of the bromthymol blue indicator from green to a deep blue. This constitutes a positive test. No growth or a green color is a negative test.
3. Record your results in the table provided.

24 Catalase and Oxidase Production

Objectives

The student will be able to:

1. describe the appearance of a positive catalase and/or oxidase test.
2. write general word equations illustrating a positive catalase and/or oxidase test.
3. name the chemical reagents used to test for the presence of catalase and/or oxidase.
4. name the two major genera of aerobic organisms which are catalase negative.
5. name the specific cytochrome involved in the oxidase test.
6. indicate a practical use for the results of the catalase and oxidase tests.

Catalase and **oxidase** are enzymes that are related to an organism's ability to utilize oxygen and are widely used as characteristics in classification of bacteria.

Catalase degrades hydrogen peroxide (H_2O_2) formed during metabolism into oxygen and water. It is not generally found in anaerobic organisms nor in certain species that are microaerophilic. This lack of catalase in anaerobes partially explains why oxygen is poisonous to them. When oxygen is available, aerobic organisms use it as a final electron and hydrogen acceptor, forming hydrogen peroxide as a result; however, anaerobes cannot degrade hydrogen peroxide any further. Thus hydrogen peroxide accumulates and poisons anaerobes because of increased concentration. It is highly useful in distinguishing morphologically similar genera of bacteria: *Staphylococcus* (+) and *Streptococcus* (–); *Bacillus* (+) and *Clostridium* (–).

Oxidase is an enzyme involving the transfer of electrons in the cytochrome chain. Kovac's oxidase test distinguishes major subgroups of chemoheterotrophs by revealing the absence or presence of cytochromes of the *c* type in the respiratory transport chain. An aerobic organism with a *c* type cytochrome can oxidize certain amines to form colored products.

A. Catalase

Materials

1. 18- to 24-hour cultures of *Enterococcus faecalis* and *Staphylococcus aureus*
2. 1 tryptic soy agar deep, sterile
3. 1 sterile Petri plate
4. 3% H_2O_2 solution

Procedure

Period 1

1. Pour one Petri plate of tryptic soy agar and allow it to solidify.
2. Using a marking pen, divide the bottom of the plate into three equal segments (see Figure 15-1).
3. Streak-inoculate one segment of the plate with *Enterococcus faecalis* and another segment with *Staphylococcus aureus.* Leave the third segment as an uninoculated control. Each streak should be about 1 cm long near the center of the segment.
4. Incubate the plate at 37°C for 48 hours.

Observations

Period 2

1. After incubation for 48 hours, place a small amount of growth from an inoculated segment on a clean glass slide. With the low power objective of the microscope focused on the cell mass, add a drop of 3% H_2O_2. The presence of gas bubbles means that the organism produces catalase and the test, therefore, is considered positive. No gas bubbles means the organism is catalase-negative.
2. Repeat step No. 1 for each organism.
3. An alternate method for catalase determination is to flood the growth on each segment of the plate with the H_2O_2 solution. If catalase is present, a trail of bubbles will arise from the growth. As before, no gas bubbles indicates the organism is catalase-negative.
4. Record your results in the table provided.

B. Oxidase

Materials

1. 18- to 24-hour cultures of *Pseudomonas aeruginosa* and *Escherichia coli*
2. Filter paper disks 5–10 cm in diameter
3. 1% aqueous tetramethyl-para-phenylenediamine HCl (Kovac's oxdidase reagent)
4. Sterile applicator sticks

Procedure and Observations

Period 1

1. Place a few drops of Kovac's oxidase reagent on a filter paper disc. *Immediately* streak a **HEAVY** mass of cells over the moist area using a sterile applicator stick (a nichrome wire loop will give a false positive reaction).

   CAUTION!

   a. A massive amount of cells must be used. A visible mass must be present on the end of the stick.
   b. Pigmented organisms frequently present a problem. Reduce the amount of cells to minimize the pigment effect.
   c. Discard the stick and filter paper in a to-be-sterilized container.

2. A change in color of the cells along the streak to a deep purple to almost a black is an oxidase-positive reaction. The color change will occur within moments. No apparent change in color from that of the reagent is a negative oxidase test.
3. Record your results in the table provided.

25 Nitrate Respiration

Objectives

The student will be able to:

1. name the specific enzyme resulting in the reduction of nitrate to nitrite.
2. define *denitrification*.
3. write the general word equations for nitrate reduction and denitrification.
4. name the chemical reagents used to test for nitrate reduction.
5. describe the appearance of a positive test for nitrate reduction.

Under **anaerobic** or **very low oxygen** conditions, many aerobic organisms are able to divert electrons from the electron transport chain and reduce the nitrogen atom in nitrate (NO_3^-) to nitrite (NO_2^-), with the latter molecule accumulating in the environment. Nitrate thus takes the place of oxygen as a terminal electron acceptor, and the process is called **nitrate respiration**. Organisms capable of carrying out this reaction possess the enzyme nitrate reductase.

Some organisms have the additional capability of reducing the nitrite, if they produce the enzyme nitrite reductase. The nitrogen atom is further reduced to nitrous oxide (N_2O) and to nitrogen gas (N_2) in the process called **denitrification**; the nitrogen atom may also be assimilated as ammonia (NH_3) into amino acids. Denitrification is sometimes exploited in tertiary waste treatment as a means of removing nitrate from treated sewage.

Nitrate respiration is determined by placing an organism in a broth medium containing nitrate under anaerobic conditions, then measuring the accumulation of nitrite. Denitrification can be determined in the same medium by including an inverted Durham tube and looking for accumulated gas after incubation. If nitrite is not found, then a determination of nitrate must be made to see if there is any present. If there is no nitrate, then all of it has been converted to gas or assimilated. If there is no gas and no nitrite, and nitrate is present, then it has not been utilized (Table 25-1).

Table 25-1

Analysis of Nitrate Utilization

Nitrate	Nitrite	Gas	Interpretation
+	–	–	no nitrate used
+	+	–	nitrate respiration
–	–	–	nitrate respiration (unlikely)
–	+	+	denitrification
–	–	+	denitrification
+	+	+	denitrification

Materials

1. 18- to 24-hour cultures of *Pseudomonas aeruginosa*, *Bacillus subtilis,* and *Staphylococcus aureus*
2. Soil sample
3. 5 tubes nitrate broth with gas inserts, sterile
4. Sulfanilic acid reagent (Nitrite A)
5. N,N′ dimethyl-1-naphthylamine reagent (Nitrite B)
6. Zinc powder
7. 6N HCl

Procedure

Period 1

1. Inoculate one tube of nitrate broth with *Pseudomonas aeruginosa.*
2. Repeat the procedure for the other organisms.
3. Inoculate a fourth tube with a small amount of soil using a wet inoculating loop. This is intended to demonstrate the presence of denitrifying and nitrate respiring organisms in soil.
4. Keep one tube as an uninoculated control.
5. Incubate all tubes at 37°C for at least 48 hours or up to one week.

Observations

Periods 2–3

1. After the incubation period, observe first for the presence of gas in the Durham tube.
2. Then test each tube and the control for nitrite by adding 1 ml of the sulfanilic acid reagent followed by 1 ml of the N,N′ dimethyl-1-naphthylamine reagent. (*Note:* 20 drops = 1 ml). If nitrite is present, the mixture of the two reagents with the medium will become red, purple, or a maroon color. *Even a transitory color is positive.* Note: Multiple incubation periods can be used by removing a small amount of culture aseptically to a sep-

arate tube where the testing is performed.

3. A negative result for nitrite requires an additional test, since the negative outcome may be due to either the absence of nitrate reduction or to the additional reduction of the nitrite, in which case all of the nitrate appears as gas. Usually the test for nitrate is performed only if the nitrite test is negative.
4. To determine which of these possibilities is correct, a small amount (small = size of a matchhead) of powdered zinc followed by a few drops of 6N HCl are added to the nitrite negative tubes. Zinc, in the presence of H^+, chemically reduces nitrate to nitrite by the formation of hydrogen (bubbles). Therefore, if a red color appears now, it means that the nitrate remained unaltered by the bacterial growth. The red color still indicates the presence of nitrite, but the nitrate was reduced chemically to nitrite by the zinc. This would not be possible if the nitrate were not present in the broth in its original form. The color change may take 5–10 minutes. If the tube did not show nitrite initially, but shows it after reduction by zinc, then the bacterium is negative for nitrate respiration or denitrification (Table 25-1). If no color change occurs upon the addition of zinc, it means that all of the nitrate has already been reduced by the bacteria beyond the nitrite stage and is, therefore, positive for denitrification.
5. Record your results in the table provided.

26 Ammonification

Objectives

The student will be able to:

1. describe the mechanism of ammonification.
2. name the chemical reagent used to test for the presence or absence of ammonia.
3. describe the appearance of a positive test for ammonia.
4. explain under what conditions ammonification occurs.

Many bacteria in nature, especially in soil and water, are capable of dissimilating organic nitrogen in protein and other nitrogenous material. When carbon is in short supply and the nitrogen is in excess of needs, the microbes utilize the carbon of the proteins and amino acids with the resulting release of ammonia to the environment. This process is called ammonification. Usually the ammonia dissolves in the surrounding water, but if it is in great excess, it may be released as a gas. The strong smell of ammonia around cattle feed lots is an example. If carbon is in excess of needs, the ammonia is not released but is assimilated by the bacteria into their own cell material.

$$\underset{\text{alanine}}{CH3-\underset{NH_2}{\underset{|}{CH}}-COOH} \longrightarrow \underset{\text{propionic acid}}{CH_3-CH_2-COOH} + \underset{\text{ammonia}}{NH_3}$$

Materials

1. 18- to 24-hour cultures of *Pseudomonas aeruginosa* and *Bacillus subtilis*
2. Soil sample
3. 4 tubes 4 percent peptone water, sterile
4. Nessler's reagent
5. White spot plate or depression slide
6. Glass rod
7. 70 percent ethyl alcohol

Procedure

Period 1

1. Inoculate one tube of peptone water with *Pseudomonas aeruginosa.* Peptone water is used because it is high in protein and amino acids and low in other utilizable carbon sources.
2. Repeat the procedure with the culture of *Bacillus subtilis.*
3. Inoculate a third tube with a small amount of soil using a wet inoculating loop. This is intended to show the occurrence of ammonia-producing bacteria in soil.
4. Keep one tube as an uninoculated control.
5. Incubate all tubes at 37°C. Make your observations at 2, 4, and 7 days.

Observations

Periods 2–4

1. After each incubation period, test each culture and the control for ammonia production.
2. Place a drop of Nessler's reagent in the depression of a white spot plate (or a depression slide on a white paper). Flame a glass rod by dipping the rod in 70 percent ethyl alcohol and passing it through a Bunsen burner flame. Once the alcohol ignites, remove the rod from the flame and let the alcohol burn off. Dip the flamed rod in the culture and add a drop of the culture to the Nessler's reagent on the spot plate.
3. A positive test for ammonia is indicated by the appearance of a yellow, orange, or brownish color on the spot plate. A brown color indicates a greater production of ammonia than does a yellow color. No color change is a negative test for ammonia. Be sure to compare each reaction with the control. Cultures exhibiting positive results need not be reincubated. Those cultures giving a negative test should be reincubated and the test repeated after the next incubation period.
3. Repeat steps 2 and 3 for each culture and the control.
4. Record your results in the table provided.

27
Urea Hydrolysis

Objectives

The student will be able to:

1. name the specific enzyme resulting in the hydrolysis of urea.
2. write a general word equation illustrating a positive urea hydrolysis test.
3. name the substance that must accumulate to give a positive urea hydrolysis test.
4. describe the appearance of a positive urea hydrolysis test.
5. explain the mechanism of urea hydrolysis and the basis of a positive test.

Urea, $NH_2 - \overset{\overset{\displaystyle O}{\|}}{C} - NH_2$, is an organic compound that is rich in nitrogen and low in carbon. Some microorganisms produce the enzyme **urease**, which splits urea into ammonia and carbon dioxide. The accumulation of sufficient ammonia creates an alkaline environment (pH of 8.1 or more), which causes the phenol red indicator in the broth to turn a red or cerise color. Organisms that dissimilate urea rapidly are sometimes of medical importance—such as the enteric bacterium, *Proteus.* Other organisms can dissimilate urea but are usually much slower.

Materials

1. 24-hour nutrient broth cultures of *Proteus vulgaris* and *Escherichia coli*
2. 3 tubes urea broth, sterile

Procedure

Period 1

1. Inoculate one tube of urea broth with 2 loopfuls of *Proteus vulgaris.*
2. Repeat the procedure with the culture of *Escherichia coli.*
3. Keep one tube as an uninoculated control.

4. Incubate all tubes at 37°C. Make your observations at 2 hours and again at 24 hours.

Observations

Periods 2–3

1. Observations of urea hydrolysis are made directly. The appearance of a red or cerise color in the urea broth is a positive test for urea hydrolysis, while a yellow or orange-red (unchanged) color indicates a negative test. Be certain to compare the color with the control tube. If negative, reincubate for a week.
2. Record your results in the table provided.

UNIT V

Environmental Stress

The environmental stresses to which organisms have been exposed since life first evolved cover nearly every aspect of physics, chemistry, and even biology—radiation, temperature, osmolytes, inorganic chemicals, pressure, oxygen, organic compounds, antibiotics, and other organisms. Microorganisms have evolved mechanisms to cope with almost every extreme of these stresses. Some archaebacteria have been reported to have an optimum growth temperature in excess of 100°C; other bacteria have been reported to grow slowly at temperatures approaching –20°C. Some organisms grow at pH 1.0 and some up to pH 13.0. Some bacteria grow in saturated NaCl, while others tolerate that condition for a long period of time. Some have adopted an alternate cell type (endospore) in response to lowered nutritional availability, which also confers extreme heat, radiation, and chemical resistance, creating a problem in the canning of food as a result. Many bacteria have evolved a system of plasmids (DNA) which may code for specialized enzymes that confer resistance to heavy metals, chemicals, and, particularly, antibiotics. Antibiotic resistance is currently an important medical problem, since these plasmids can often be transferred to other species of the same genus and even to other genera of bacteria. Plasmids are also being exploited in genetic engineering. As a result of these many adaptations to extreme conditions, there are very few environments on the face of the earth that do not have a complement of microorganisms—with the possible exception of fluid lava.

This unit presents a selection of exercises intended to illustrate some of the adaptations bacteria have made through their evolution.

28
Temperature

Objectives

The student will be able to:

1. define psychrophile, mesophile, and thermophile.
2. define maximum, minimum, and optimum temperature of growth.
3. determine the growth temperature range and approximate optimum growth temperature of a bacterial culture.

An organism's response to heat in the environment is related to the organism's normal habitat. Generally speaking, cold environment organisms do not grow well or even at all at temperatures above 20°C. These organisms are called **psychrophiles** (*psychro* [G]-cold/phile [G]-loving). Other organisms grow best at temperatures above 45°C and rarely grow below 40°C. These are called **thermophiles** (*thermo* [G]-hot). Still other organisms grow best in the range 20°C to 45°C, and these are called **mesophiles** (*meso* [G]-middle). These terms are used to describe the range in which the optimum growth temperature is found, not necessarily where growth ceases. Many of these organisms grow slowly outside the indicated range, and the term **facultative** may be used to indicate that. A special group of mesophiles important in food spoilage and growing at 0°–5°C are called **psychrotrophs.**

Minimum and **maximum growth** temperatures are temperatures below or above which a given organismal strain ceases to reproduce and, in fact, may be killed. Death is commonly the case just above the maximum temperature of growth. Temperatures below the minimum are often not lethal and may be preservative to varying degrees. The **optimum** temperature of growth is generally the temperature at which division is most rapid. However, the production of some metabolic product may be optimal at a temperature other than the optimum for division. For example, more lactic acid might be produced at a temperature below the growth optimum than at the optimum itself, even though division is slower at the lower temperature. Advantage is taken of this in cheese making.

The extremes of temperature at which microorganisms can grow, especially prokaryotes, is truly amazing: from about –12°C in super-cooled foods to 110°C in undersea thermal vents (e.g., Galapagos trench). An archaebacterium has been reported from a thermal vent with an optimum of 105°C and with maximum and minimum temperatures of 110°C and 90°C, respectively. In most cases it is believed that temperature extremes from the optimum affect enzyme formation or function (by denaturation), resulting in no metabolism or no reproduction. In some cases the cell membrane structure may be altered, either solidifying (freezing) or liquefying (melting), resulting in transport or membrane failure.

This exercise illustrates a visual method for determining the range and approximate optimum growth temperature for several organisms. An optical absorbance method using a spectrophotometer is used for more refined studies.

Materials (groups of 4)

1. One 18- to 24-hour-old tryptic soy broth culture tube of each of *Pseudomonas fluorescens*, *Escherichia coli*, and *Bacillus stearothermophilus*
2. One 18- to 24-hour-old marine broth culture of *Vibrio marinus*
3. 19 tryptic soy broth tubes, sterile
4. 7 marine broth tubes, sterile
5. Incubators or water baths at each temperature

Procedure

Period 1

1. One student should label 6 tubes of marine broth with the name *V. marinus* and one of 6 incubation temperatures: 5°, 20°, 30°, 40°, 50°, and 60°.
2. Each remaining student in the group should label 6 tryptic soy broth tubes similarly with the name of one of the organisms provided and a temperature of incubation. One tube is reserved as a no-growth control for comparison. *Note:* Use an indelible marking pen if water baths are used as incubators.
3. Inoculate the labeled tubes with *one* loopful of the assigned organism.
4. As quickly as possible, incubate each tube at the appropriate incubation temperature. Don't delay. Even a moderate time at room temperature may permit cell division.
5. Observe after 2 and 7 days of incubation.

Observations

Period 2

1. At 48 hours, handle the tubes carefully, looking at each one for a small button of growth. Do this by holding the tube over your head and looking up from below. Record as positive (+) for growth if a button is present.
2. Shake each tube gently and compare the turbidity against the control tube. Record as positive (+) in the results table *only* if growth has occurred. If there is no growth, leave blank.
3. Return tubes to the incubator as quickly as possible.

Period 7

1. At 7 days, again look for a button on the bottom of the tube. If the button is no larger than on Day 2, record as negative (0) for growth.
2. For each organism, shake each tube gently to suspend cells and compare against the uninoculated control tube. Find *the one* tube for that organism of *all* the temperatures of incubation that shows the *most* growth (i.e., the tube with the most turbidity). This tube becomes ++++ (i.e., shows the maximum amount of growth).
3. Now compare each of the other tubes of the same organism with the tube rated ++++, this time recording growth (turbidity) as 0 (no growth), +, ++, +++, or ++++.
4. Repeat steps 2 and 3 for each organism used.

Interpretation

The results of this experiment do not lend themselves to an accurate determination of optimum temperature but usually will provide a range. The tube showing 4+ will be near the optimum. In cases where two tubes are 4+, the optimum may lie between them. Temperature gradient bars are commonly used to determine the optimum more accurately. An incubation temperature at which growth does not occur, but it does occur at the next temperature below, can be used as an estimate of the maximum growth temperature. Minimum growth temperature is more difficult to determine and requires prolonged incubation. The growth temperature ranges of the psychrophiles, mesophiles, and thermophiles usually can be seen from these results.

29
Osmotic Pressure

Objectives

The student will be able to:

1. distinguish between hypotonic, isotonic, and hypertonic solutions.
2. define plasmolysis and plasmoptysis.
3. define halophile and name a halophilic bacterial genus belonging to the archaebacterial group
4. define osmophile and xerophile.

The presence of ions and organic substances in solution binds a certain amount of water and creates an **osmotic pressure** across a semipermeable membrane, such as that around the cytoplasm of a cell. If the amount of "free water" (i.e., not bound to the solute and sometimes called the aqueous activity or a_W) differs on each side of the membrane, there is a net movement of water from the side with more "free water" to the side with less "free water"—that is, from the side with fewer ions or organic matter to the side with more. This water movement continues until the osmotic pressure is equal on both sides, a condition called **isotonic**. If there is a higher free-water concentration outside the cell, the solution is **hypotonic**, and water moves into the cell, causing it to swell and possibly to burst if the osmotic pressure difference is great enough. This imbibing of water and consequent swelling is called **plasmoptysis**. The reverse process, water leaving the cell to a more concentrated solute or **hypertonic** solution outside the cell, causes the cell membrane to shrivel and shrink, a process called **plasmolysis** (Figure 29-1).

Many organisms are able to modify their cell chemistry in such a way as to grow in the presence of solutes, sometimes in rather high concentration, but do not require the solute for growth. An example is sodium chloride (NaCl). Organisms growing in the presence of NaCl but not requiring it are called **halotolerant** (*halo* [G]-salt). Some organisms have adapted to the presence of high salt concentration in such a way that they require the presence of the solute to grow at all. These organisms

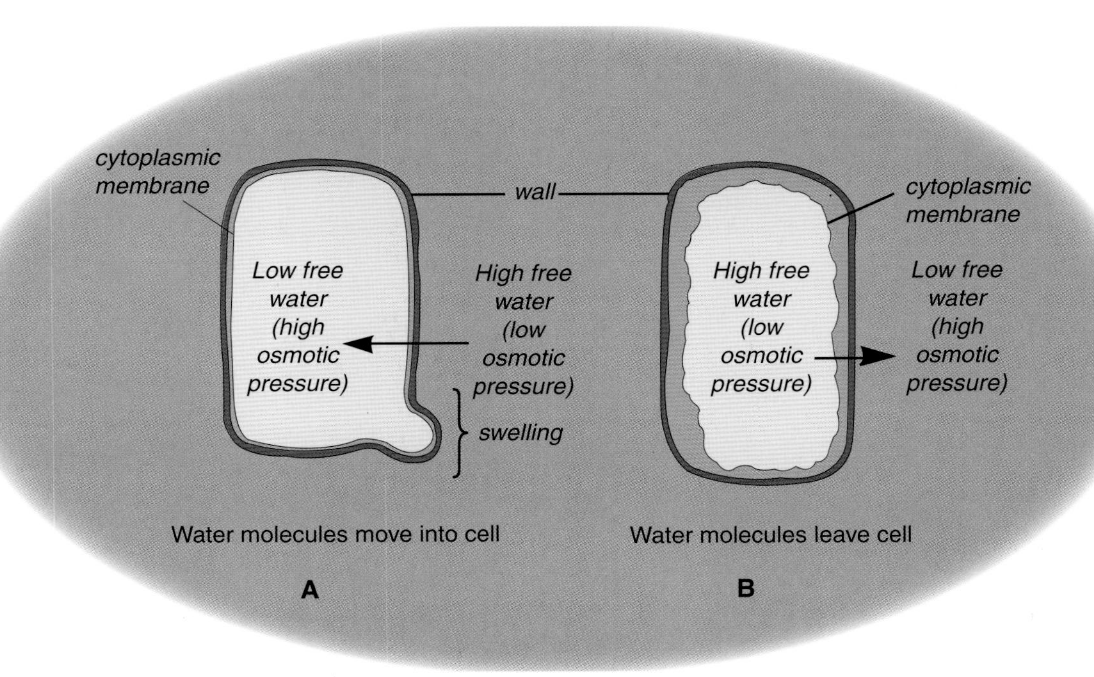

Figure 29-1

Osmotic pressure: (A) Plasmoptysis (B) Plasmolysis

are called **halophiles**. Many marine bacteria require salt at fairly low levels. Such organisms are moderate halophiles. The **extreme halophiles**, such as the archaebacterial genus *Halobacterium*, require salt in excess of 12% and grow well at NaCl saturation (27% w/w) in salt lakes, or salterns (salt evaporation ponds), and evaporation pools along the ocean shore. These are usually bacteria. Sugar and dry conditions also exert a strong osmotic pressure to which organisms have adapted. In sugars and syrups, yeasts often grow and are called **osmophiles** (*osmo* [G]-pushing). On dehydrated materials, especially foods, molds can grow; these are called **xerophiles** (*xero* [G]-dry).

This exercise is designed to demonstrate one type of osmotic pressure tolerance, resistance to sodium chloride.

Materials (pairs)

1. Cultures of *Escherichia coli*, *Staphylococcus aureus*, *Saccharomyces cerevisiae*, and *Halobacterium salinarium*
2. 5 prepoured Petri plates of different concentrations of salt agar, sterile
   a. 0.5% NaCl
   b. 5% NaCl
   c. 10% NaCl
   d. 15% NaCl
   e. 20% NaCl

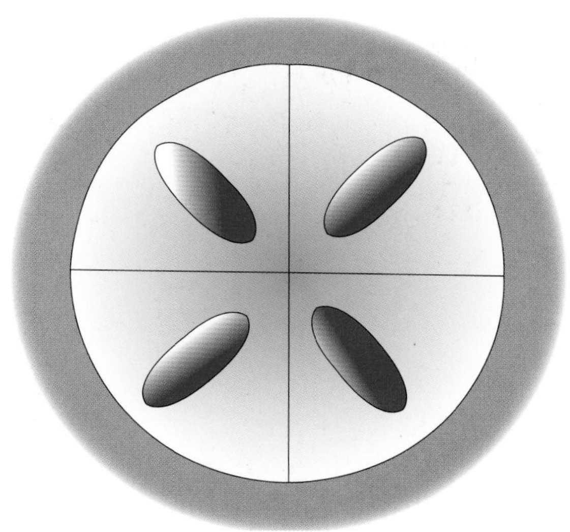

Figure 29-2

Marked and streaked plate

Procedure

Period 1

1. With a marking pen mark the salt concentration on the underside of each Petri plate and divide it into quadrants as illustrated in Figure 29-2. Label each section with the name of one of the organisms.
2. Using a sterile loop, streak one organism on its quadrant beginning near the center and toward the side. Do not get too close to the center or the side (Figure 29-2).
3. Streak the same organism on each plate, going back to the stock slant each time.
4. Repeat for the other organisms on the appropriate quadrant.
5. Incubate all plates at 32°C for 48 hours.

Observations

Periods 2–3

1. Observe the plates for growth at 48 hours and again at one week.
2. Record the results on the report form as follows:
   Find the plate where a particular species has the greatest amount of growth. That is the control plate for that organism and is scored ++++. Compare each other plate for that organism only and record the amount of growth as 0 (none), +, ++, +++, or ++++. Maximum growth should not occur at the same salt concentration for all the organisms.
3. Record your results on the form provided.

30
pH

Objectives

The student will be able to:

1. make a generalization about what kinds of organisms grow at low pH and at high pH.
2. name two species of bacteria, one an archaebacterium, able to grow at pH <2.0.

Among the environmental stresses an organism may encounter, pH is of considerable importance. Each organism has an **optimum pH** for growth. Most organisms grow well, if not best, at a pH between 6.5 and 7.5, although many grow well, even optimally, outside this range. The majority of microbes grow best around a pH of 7.0, but some, called **acidophiles**, do very well in the 3–4 range (e.g., the lactic acid bacteria). Some members of the bacterial genus *Thiobacillus* and the archaebacterial genus *Sulfolobus* not only grow best at pH <2.0 but generate their own acid; they are commonly found in acid mine wastes. *Sulfolobus* additionally grows at very high temperatures. Some of these bacteria will grow at pH 1.0 or less. Yeasts and molds are generally very acid tolerant and grow quite well at pH 3.0 or a little lower. Molds and yeasts generally will grow at lower pH than bacteria. This is the reason acid foods such as fruit juices are usually spoiled by yeasts and molds rather than bacteria. Media for the selective isolation of yeasts and molds takes advantage of this by lowering the pH and suppressing bacteria. Some algae, but not cyanobacteria, grow at or near pH 2.0.

At higher pH values, a similar situation exists. Organisms grow as high as pH 12 and are called **alkaliphiles**. These include some bacteria, cyanobacteria, algae, protozoa, and a few insects. Alkali lakes are found all over the world in arid climates but have not been studied as extensively as acid environments. It is possible that some organisms grow at pH values above 12, but such environments do not occur naturally. Many of the alkaliphiles also require Na^+, some in excess of 15%. *Vibrio cholerae*, the causative agent of cholera, will grow readily at pH 9.0–9.6 while most other enteric organisms do not. Selective media for *V. cholerae* are often adjusted to pH 9.0 for this reason.

Materials (groups of 5)

1. Each student will be assigned a broth culture of one of the following:
   Bacillus sp. pH 9.5–10
   Saccharomyces cerevisiae pH 7.0
   Staphylococcus aureus pH 7.0
   Escherichia coli pH 7.0
   Enterococcus faecalis pH 7.0
2. Each student will have one sterile tube of tryptic soy broth at each pH of 4, 5, 6, 7, 8, 9, and 10

Procedure

Period 1

1. Label each tube as you pick it up. Inoculate a loopful of your assigned organism into each labeled tube of the pH set. Be sure to sterilize the loop between each tube so that no acid or base is transferred to the stock tube and the inoculum size is the same.
2. One uninoculated tube of each pH should be saved for the class as a control.
3. Incubate the tubes and controls at 35°–37°C except for *S. cerevisiae*, which is incubated at 30°C.

Observations

Periods 2–3

1. At 48 hours, examine each tube for turbidity, comparing it to the control for that pH.
2. Record the results on the report form as 0 (no growth), +, ++, +++, or ++++. This last would be the pH of greatest growth.
3. Reincubate the tubes and repeat the observations at 7 days.
4. Collect data for the other organisms from the other members of your group.

31 Ultraviolet Light

Objectives

The student will be able to:

1. perform a simple demonstration of the effects of ultraviolet light.
2. define *photoreactivation*, *light repair*, and *dark repair*.
3. explain how ultraviolet light might cause mutation.
4. explain why the height of the ultraviolet lamps in this exercise is important.

Ultraviolet light is widely used to disinfect or sterilize air in hospital air circulation systems and surgical tools and has a number of applications in the food industry for control of airborne mold spores, bacteria, and bacteriophages around packaging equipment, fermentation vats, and other places. Ultraviolet light is highly germicidal between 130 and 400 nm but is most effective at 256 nm. Double bonds in the purines and pyrimidines of DNA absorb strongly near this wavelength, resulting in rupture and reformation in unusual configurations. Adjacent thymines, in particular, form a **thymine dimer** which prevents DNA duplication, thus causing the death of the cell. Many microbes have enzyme systems that can repair such damage in some cases. One system requires the presence of visible light in the 420–540 nm range. This is sometimes called **photoreactivation** or the light repair system and results from a photoreactivating enzyme called DNA-photolyase (see *Science* 266:1954, 23 Dec 1994). Some organisms possess a **dark repair** enzyme system which operates in the absence of light. Out of all of the cells affected, only a few can perform these repairs and allow growth. Some of the surviving cells may carry **mutations** as a result of changes in the DNA codons of the cell. These express themselves in different ways.

Ultraviolet light has a number of drawbacks as a germicidal agent. It has very poor penetrating power and is effectively blocked out (absorbed) by clear glass, plastics, thin films of water, and any opaque material. Thus it is useful only where the organisms are directly exposed to the light waves. Ultraviolet light also can cause severe eye and skin damage to workers, even when reflected from surfaces. Direct exposure to the skin and eyes must be limited. Reflected light is also dangerous.

This exercise is intended to illustrate the effects of ultraviolet light and some of its limitations.

Materials (per pair in a group of 3 pairs)

1. *Staphylococcus aureus* or *Serratia marcescens* broth culture
2. 1 sterile cotton swab
3. 5 Petri plates, sterile
4. Nutrient agar, sterile
5. 3 ultraviolet lamps (may be one set for whole class)
6. 3 sheets of paper (5 x 8 inches)

Procedure (a group of 3 pairs)

Period 1

1. Place three germicidal lamps well separated from each other at 15, 30, and 60 cm, respectively, above the bench top.
2. Work in a group of three pairs for each organism. Each pair in the group uses a different lamp height.
3. Each pair should pour nutrient agar into 5 Petri plates and allow them to solidify.
4. Each pair should set one plate aside as a control.
5. Wet a large sterile cotton swab in a broth culture of the assigned organism. Press out the excess fluid on the inside wall of the tube and spread the swab evenly over the surface of one of the plates. Repeat this step for each plate.
6. Each pair should then cut a pattern of choice in three 5 x 8 papers slightly larger than the size of the Petri plate. Each pattern should expose about half of the Petri plate (Figure 31-1).
7. The remaining four Petri plates are placed cover up under the assigned germicidal lamp. Remove the covers of three of the plates and replace with the paper cutouts. The fourth plate is exposed with the cover in place (Figure 31-1).

CAUTION

Expose your hands to the light only briefly and be careful not to let the UV light reflect into your eyes.

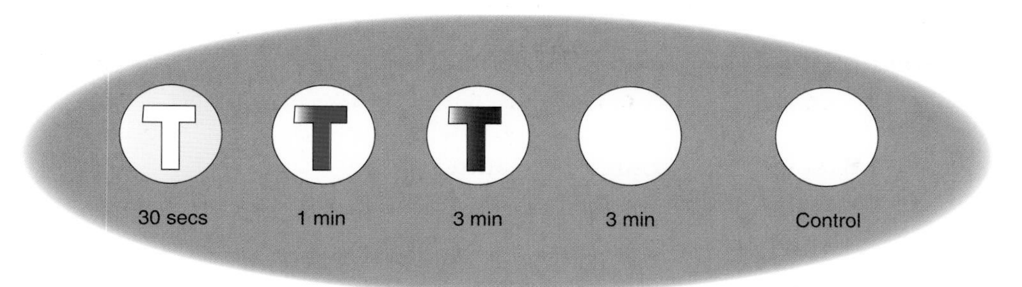

Figure 31-1

Ultraviolet light exposure times

8. Expose one paper-covered plate for 30 seconds, one for 1 minute, and one for 3 minutes. The covered plate is exposed for 3 minutes. Label the plate covers with your name, lamp height, and time of exposure.
9. The plates of *S. marcescens* **only** are placed in a dark box or wrapped with aluminum foil to exclude all light.
10. Discard the paper shields in an autoclave bag, **NOT** the trash.
11. Incubate all plates at 30°–32°C.

Observations

Period 2

1. After 24–48 hours of incubation, make drawings of each plate in the space provided on the Results and Observations form showing the relative amount of growth as compared to the control plate.
2. Count surviving colonies in the exposed area of each plate and calculate the relative survival (if any) as a function of time of exposure and lamp height.
3. Make a note of all colonies that are white or sectored with white. These represent pigmentation mutations. A mutation rate can be calculated by counting all white or sectored colonies in the exposed area and dividing by the total number of all colonies. To determine the rate specifically due to ultraviolet light, the spontaneous mutation rate must be known. This exercise will not provide the spontaneous rate.

32 Bacterial Motility

Objectives

The student will be able to:

1. determine whether a bacterial culture is motile or non-motile.
2. recognize Brownian movement.
3. prepare a wet mount for determining motility.
4. inoculate a motility medium and interpret the results (may be omitted).

An important characteristic of some microorganisms is the ability to move, thus enhancing an organism's opportunity to acquire food from areas of fresh supply or escape from harmful substances (i.e., **chemotaxis**). Eukaryotic cells have a number of means of locomotion: complex flagella (flagellates) or cilia (ciliates) which

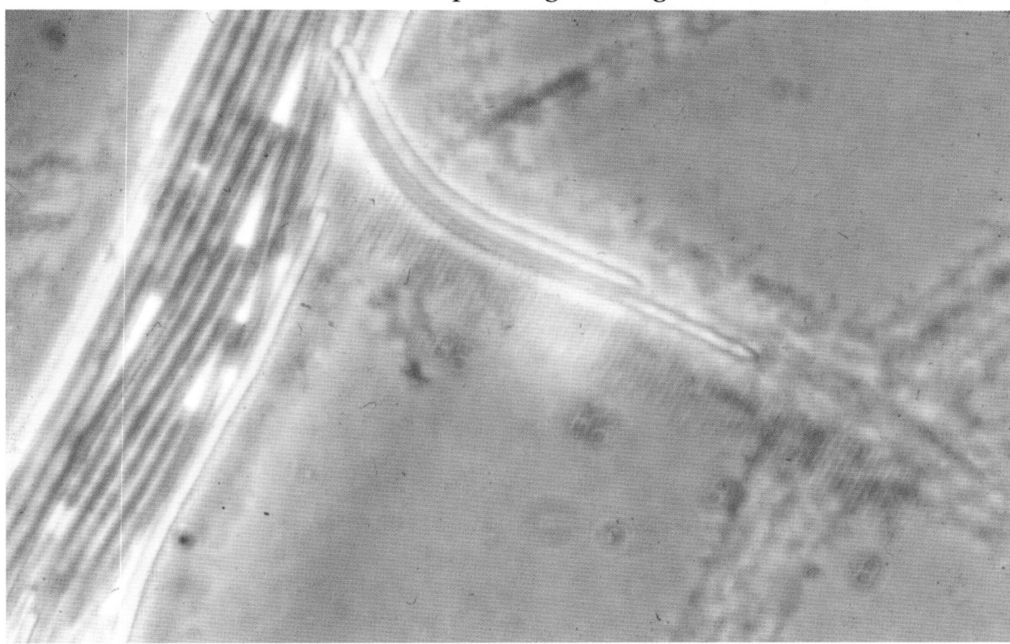

Figure 32-1

Cyanobacteria—*Oscillatoria*, gliding motility. Note slime tracks.

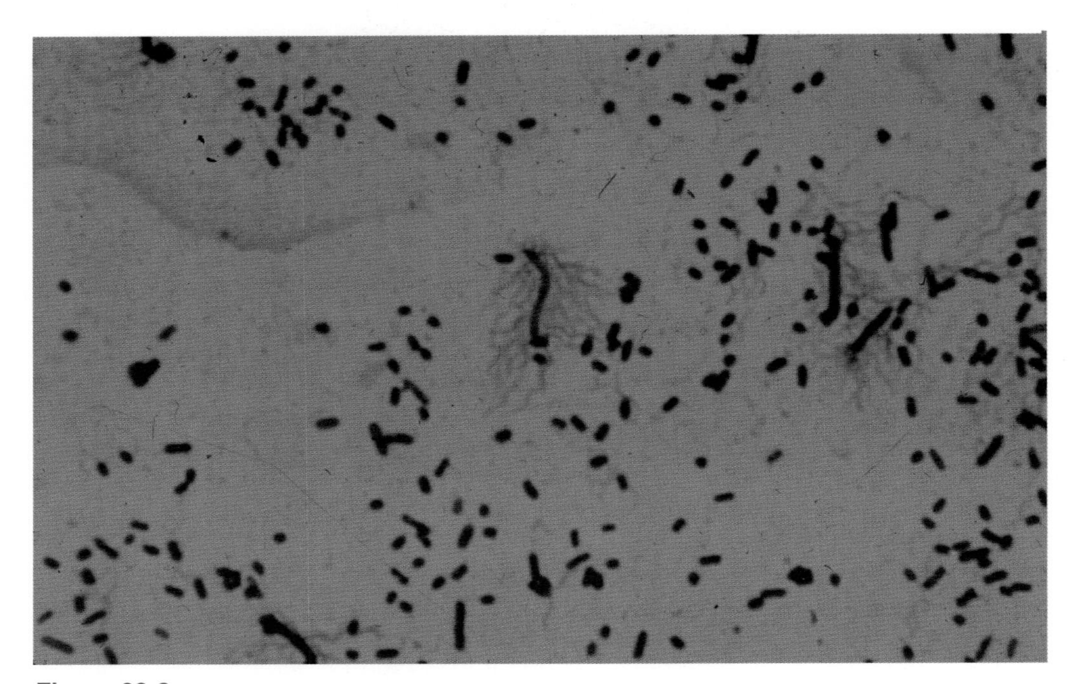

Figure 32-2

Eubacterium showing peritrichous flagellation

are large enough to be seen through the microscope, amoeboid movement (amoeba), and gliding (diatoms) types of motility are found. Prokaryotes move by gliding (Figure 32-1) or with flagella (Figure 32-2), of a type too small to be seen through the microscope and not at all like the flagella of eukaryotes. Inability to move is found in both groups. Nonmotile bacteria frequently exhibit Brownian movement, a random movement with a different direction each time resulting from bombardment of the cell with water molecules. Only the small bacteria show this. Larger bacteria and eukaryotic cells are too large to show Brownian movement.

Motility of all types can be determined by directly observing live cells in a wet mount preparation under the microscope or, in the case of bacteria, by inoculating a very soft agar medium and looking for growth beyond the stab line. The growth results from the ability of motile cells to move through the soft medium which nonmotile bacteria cannot do. The soft agar technique is not as sensitive as the wet mount method.

The location and number of flagella on a cell also provide useful information, especially in identification and taxonomy. Because bacterial flagella are so small, they must be specially stained to make them visible or observed with the aid of an electron microscope.

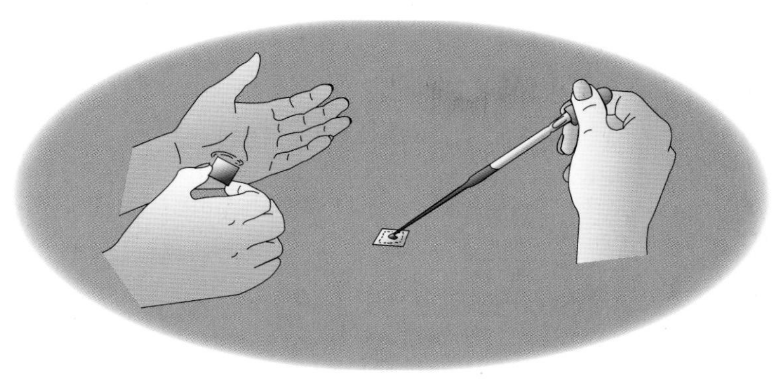

Figure 32-3

Preparing a wet mount

Materials

Part A

1. 12- to 18-hour slant culture of *Pseudomonas aeruginosa*, *Bacillus subtilis*, and *Staphylococcus aureus*
2. Depression or plane microscope slides
3. Coverslips (22 x 22 mm)
4. Vaseline
5. Toothpicks

Part B (may be omitted)

1. 12- to 18-hour slant culture of *Pseudomonas aeruginosa*, *Bacillus subtilis*, and *Staphylococcus aureus*
2. 4 tubes motility test medium

Procedure

Period 1

A. Wet Mount Preparation

1. Obtain a depression slide (or a plane slide for phase microscopy).
2. Using your hand or a toothpick, ridge each edge of a 22 x 22 mm coverslip with a *small* amount of Vaseline. Do NOT make the ridge too thick (Figure 32-3).
3. With a 3–5-mm-diameter loop, place a drop of freshly boiled distilled water on the center of the coverslip with the Vaseline side up.
4. Emulsify a small amount of a young culture in the drop. It should be *faintly* turbid. (Note: Use 12- to 18-hour-old cultures, since older cultures may lose their motility.)
5. Invert the depression slide and place it over the coverslip, pressing down lightly to make a seal. Quickly turn the entire preparation right side (coverslip) up.
6. Place the slide on the microscope stage and reduce the light. Using the low power objective, locate the edge of the drop and

center it across the field of the microscope.

7. Switch to the high-dry objective and refocus on the edge of the drop.
8. Place a drop of oil on the coverslip and *carefully* switch to the oil immersion objective, refocusing on the edge of the drop. Adjust the condenser and diaphragm for optimum contrast and move over the drop.

Observations

Period 1

1. Look for cells and motility or Brownian movement. Cells are often hard to see because of their low contrast. This is best done with a phase contrast microscope.
2. In young motile cultures, most cells will be motile while in old cultures, only a few cells may be motile. Some cultures are only sluggishly motile even when young. Generally, a cell that moves 3 or 4 times its own length in one direction is considered motile. If none of the cells move this distance, it is probably Brownian movement due to bombardment by water molecules. In Brownian movement, a random movement is imparted to the cells, rarely moving them more than once their own length in one direction. Patience is cautioned in observation! Examine a

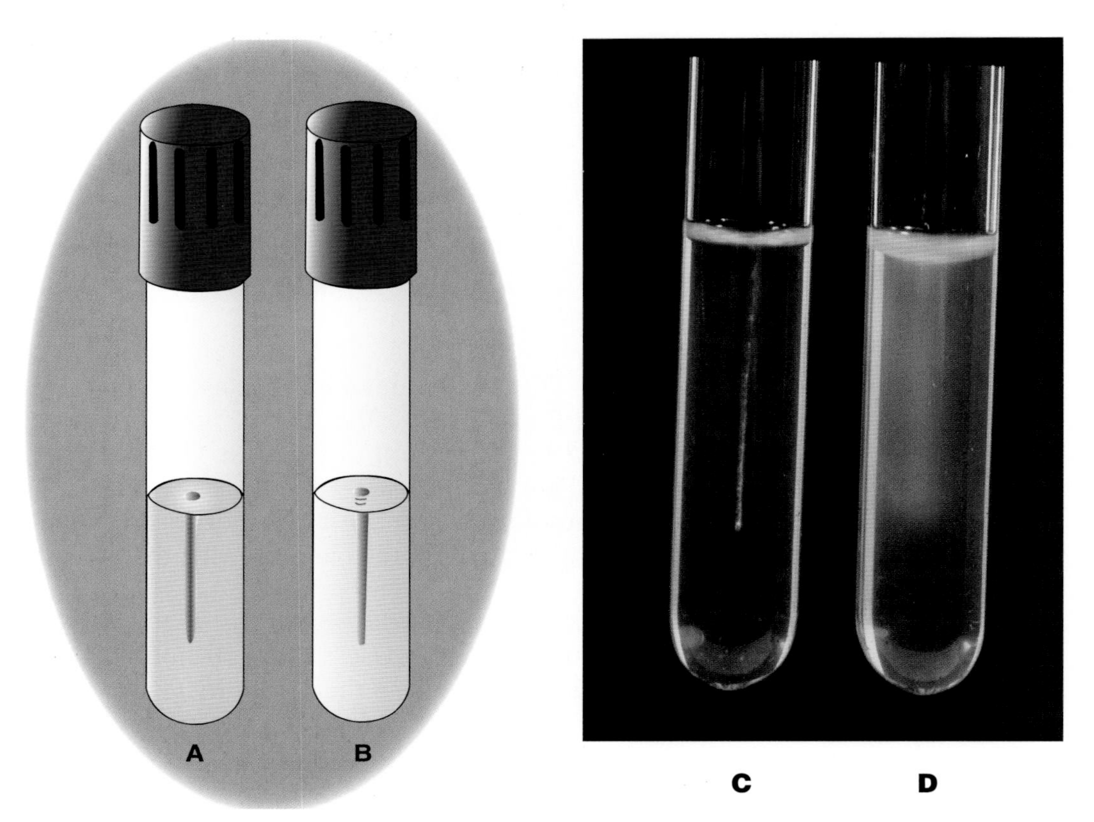

Figure 32-4

Motility test medium. (A & D) Motile organisms. (B & C) Nonmotile organisms.

field for some time before moving on.

3. Record your observations in the table provided. Describe the type of motility using such terms as darting, sedate, undulate, corkscrew, rapid, slow, only a few, most move, sluggish, etc.
4. Repeat your observation for each culture provided.

B. Soft Agar Method (may be omitted)

Period 1

1. Using a straight inoculating needle, stab a tube of motility test medium with *Pseudomonas aeruginosa*. Care should be taken to withdraw the needle along the line of the stab (Figure 32-4).
2. Repeat the procedure for each of the remaining cultures.
3. Leave one tube as an uninoculated control.
4. Incubate for 24–48 hours at the appropriate temperature for each organism.

Observations

Period 2

1. After incubation, using the control tube for comparison, observe for growth restricted to the stab line (nonmotile) or diffused into the medium (motile), which may cause cloudiness. Some motile organisms may spread throughout the tube and across the surface.
2. Record your results in the table provided. Describe how far into the agar growth has occurred and whether growth is heavy or light.

UNIT VI

Identification of Unknown Bacteria

A critical part of taxonomy, clinical microbiology, and other areas of microbiology is the need to characterize an organism to the point that it can be compared to others in order to establish a taxonomy or to identify the organism for medical or other reasons. In this unit two approaches are used: a taxonomic approach and an identification procedure. Identification is generally based on a useful subset of taxonomic characters. If available, the student may also be introduced to computers as used in taxonomy and identification. The instructor may choose to do the more formal Exercise 33, which is the basis of a taxonomic procedure as well as identification, or may choose to do one or the other parts of Exercise 34, which are widely used identification procedures in the clinical setting. In either case, the student has an opportunity to identify an unknown bacterium. If the unknown of Exercise 33 is begun earlier in conjunction with the physiological and biochemical exercises, some tests called for in Exercise 33 will already have been done.

33
Identification of an Unknown Organism

Objectives

The student will be able to:

1. identify an unknown organism using *Bergey's Manual of Systematic Bacteriology* and other diagnostic keys.
2. use the collected data with a computer program, if available.
3. explain the meaning of a similarity coefficient.

Taxonomy concerns the arranging of organisms into groups based on common characteristics. This process is also referred to as **classification**. A group of organisms so arranged is called a **taxon**. When a name is given to this taxon following specific rules, it is called **nomenclature. Identification** is the placing of an unknown organism into an established taxon on the basis of shared characters. This exercise is designed to acquaint you with the identification of unknowns in the microbiology laboratory using computers and a numerical approach. (*Note:* This exercise can be done without recourse to computers—see discussion below.) The approach is to make a number of observations and inoculate a battery of diagnostic media with an unknown organism, then read and record the results on the Data Collection Sheet provided for this exercise. Using the data on the collection sheet, your instructor will select one of several methods for identifying the unknown: a **similarity coefficient computer program TIDENT** *or* a **dichotomous key** either computer-generated or printed. Once your organism is identified, you must then consult *Bergey's Manual of Systematic Bacteriology* (Vol. 1, 2, 3, or 4), compare your data, and decide whether the identification is correct or not. The Data Collection Sheet and your analysis are then submitted to the instructor.

The **similarity coefficient** program compares the characters determined for your unknown with a set of known organisms, one character at a time and sums the number of similarities (S) and the number of differences (N)—ignoring comparisons where either organism has missing data—and calculates a simple matching coefficient as a percent similarity (% S_{sm}) as follows:

$$\%S_{sm} = \frac{S}{S + N} \times 100$$

While this program is a simple matching coefficient, the same method of calculation and matching is also used in more sophisticated taxonomy methods. The procedure here uses only 50 characters; 100 or more is preferred. After comparison with all the organisms in the data bank, the computer prints the name of the organism or organisms most closely resembling your unknown, the percent similarity, and the complete coded data. If your data are not sufficiently close to one or has the same %S with more than one organism, then a suggestion is printed for you to check your coding or reevaluate your data.

The **dichotomous key** is available as a computer program or as a printed form. A question is posed about a particular characteristic for your unknown. Respond with a yes or no from the Data Collection Sheet. You will then be directed to another question and so on until your organism has been identified. You must then verify this with *Bergey's Manual of Systematic Bacteriology* as described above.

Materials

1. An 18- to 24-hour slant culture of an unknown will be provided.
2. Your instructor will designate the incubation temperature to be used for your unknown.
3. Your instructor may choose to provide the unknown at the time the metabolic studies are begun (Exercise 15), which will permit you to compare your results with positive and negative cultures for a particular test. The media list under **Procedure** is presented with this in mind.
4. If the similarity coefficient program is used, all 50 characteristics on the Data Collection Sheet should be determined. This is a minimum for reasonable similarities.
5. If the dichotomous key is used (computer or printed form), your instructor may designate the minimum set of tests and media necessary.

Procedure

Period 1

A. When given your unknown either here or with Exercise 15:

1. Record the number of your unknown on the Data Collection Sheet.

2. Inoculate one tryptic soy agar slant and incubate it at your designated temperature for 24-48 hours. Place it at room temperature for storage. **This will serve as your reserve or stock culture in case some contamination occurs.**
3. Use your original culture for inoculations. Transfer to new media occasionally.
4. If the unknown is given with Exercise 15, **stop here**; otherwise proceed with Part B.

Period 1

B. (Parentheses refer to Data Collection Sheet characters—see relevant exercises for procedures)

1. Inoculate 4 tubes of tryptic soy broth. Incubate one tube each at 25°C, 37°C, 45°C, and 55°C for 48 hours to determine the maximum temperature at which growth occurs (21).
2. Make preliminary observations of Gram reaction (8), acid-fastness (10), cell shape (1), end of cell (4), and arrangement (5). If micrometers are available, measure Gram-stained cell dimensions, including length and width (2, 3).

Period 2

C. After 48 hours

1. If your unknown organism is a Gram-positive rod, inoculate a manganese agar slant and incubate at the designated temperature for 18 hours. Then perform the endospore stain (6).
2. Inoculate all media below not already provided with Exercises 15–28. Incubate the media at the designated temperature for your organism for the time period indicated in the relevant exercises.

Media not previously supplied

a. Streak for isolation on a plate of plate count agar and incubate for 48 hours. Determine colony size, elevation, and margin (11, 12, 13), density (17), pigmentation (14), pigment solubility (15), and fluorescence under ultraviolet light (16).
b. Growth characteristics in tryptic soy broth (18, 19, 20).
c. Oxygen requirements—pipet 1 ml of a broth culture grown at the designated growth temperature into a melted but cooled yeast extract-tryptone 0.75% agar shake, mix and allow to solidify. Incubate for 48 hours (7).
d. Motility—from a slant using wet mount or soft agar method (9)—See Exercise 32.
e. Sugar fermentation—fructose (23), galactose (24), maltose (27), mannitol (28), sorbitol (29), arabinose (30), and xylose (31)—See Exercise 21.

If not previously supplied with Exercises 15–28 perform the following:

f. Sugars—glucose (22), lactose (25), and sucrose (26)—See Exercise 21.

g. Litmus milk reactions (32, 33, 34)—See Exercise 18.
h. Hemolysis (35)—See Exercise 20.
i. Starch hydrolysis (36)—See Exercise 15.
j. Gelatin hydrolysis—Fraser's overlay method (37)—See Exercise 17.
k. Casein hydrolysis—skim milk overlay method (38)—See Exercise 17.
l. Lipid hydrolysis (39)—See Exercise 16.
m. Nitrate respiration and/or gas (40,41)—See Exercise 25.
n. Ammonia production (42)—See Exercise 26.
o. Catalase and oxidase (43, 44)—See Exercise 24.
p. Hydrogen sulfide production—Kligler iron agar (45)—See Exercise 22.
q. IMViC reactions (46, 47, 48, 49)—See Exercise 23.
r. Urea hydrolysis (50)—See Exercise 27.

Observations

Periods 2–3

1. Enter all your data on the Data Collection Sheet by circling the number designation for your unknown results. Circle "0" if the test was not done or lost.
   Note: Be careful of using "0": If other results imply a valid score, use it. For example, step C-1 above indicates only Gram positives should use manganese agar and do a spore stain. If your organism is Gram negative, Character 6 would be scored "1" (none) not "0". The fewer "0's" scored, the better the match will be. Check with your instructor if you have a question.
2. Record the character score in the column immediately to the left of the Character No. column for ease in responding to the computer or key questions.

Computer

3. Now you are ready to use the computer. Your instructor will provide you with detailed instructions for use of the particular computer. Enter your data as called for and wait for the answer.
   a. The computer program has a cutoff of 85%S for identity. If you are lower than this or more than one organism has the same %S, you will need to review, repeat tests, or recode your data. Repeat the computer run if necessary.
   b. Go to step 6 below.

Dichotomous Key

4. If the dichotomous key is used, it is available in two forms: as a computer program or a printed sheet. In the former, your instructor will provide you with detailed instructions in use of the particular computer. Questions are displayed one at a time. Respond by answering yes or no. The program then takes you to the next question on that branch. Continue responding until

your unknown is identified or you are told an error has occurred, in which case you will need to review, repeat tests, or recode your data. Go to Step 6 below.

5. If the printed dichotomous key is used, simply follow the questions and branches as indicated until your organism is identified or you come to an error, in which case you will need to review, repeat tests, or recode your data. Go to Step 6 below.

Report

6. Take your identified unknown and all the data you have collected and find your unknown in *Bergey's Manual of Systematic Bacteriology* (Vol. 1, 2, 3, or 4). Compare with your unknown data and decide if this is indeed the correct identification. Write a short report on how the Bergey's data and your data differ or agree and how confident you are in the identification. A very important part of this report is to indicate *which tests could be used to help further clarify the identification.* If these reference books are not available, your instructor will provide additional material.

34 Identification with a Miniaturized, Rapid Biochemical System

Objectives

The student will be able to:

1. identify an unknown member of the *Enterobacteriaceae* using the API 20E system or the ENTEROTUBE II system.
2. describe the advantages of the multitest system over the conventional methods.
3. explain the difference between identification and taxonomy as used in this exercise.
4. be able to answer questions about the introduction to Exercise 33 as well.

The introductory material of Exercise 33 should be read as part of this exercise even if Exercise 33 was not done. In clinical microbiology, microbes must be identified quickly and accurately with a minimum of cost. It is not necessary to know how the organism fits in the taxonomy of all other organisms but only which one it is. To do this, tests are often weighted (i.e., the possession of a single character has great importance in recognizing what it might be). Dichotomous keys and probabilistic models are derived from taxonomic characters and used as aids in identification. Traditionally this has involved inoculation of many media, as exemplified in Exercise 33 and the physiological tests done earlier. This is expensive and time consuming, and recent developments in clinical laboratory practice have centered around reducing the time and cost involved.

A number of commercial kits have been developed involving multiple biochemical tests done at one time, thus leading to the name "multitest systems". These include the API, ENTEROTUBE II, and Micro-ID systems, among others. Originally developed for the *Enterobacteriaceae*, perhaps the most commonly encountered clinical group, kits are now available for other bacterial groups as well. Patterns of results are compared to experience with known strains from many laboratories over a long period of time. This bank of data is computerized or supplied as a comprehensive summary of results by the manufacturer of the kit. By this comparison, a tentative identification can be made in much less time than by conventional methods. Several precautions must be noted. First, the instructions for the kit must be followed **exactly** and a **pure culture** is absolutely essential. Second, use of the kit with organisms other than the ones specifically designated must be done with caution. Third, results with the kits often must be correlated with other identifying characteristics, such as specimen source and colony morphology on both differential and selective media, and oxidase reaction, before an identification can be made. Results sometimes must be confirmed using other techniques such as serology or other biochemical tests.

This exercise is intended to introduce the student to the use of a multitest kit for identification of an unknown from the *Enterobacteriaceae* using one of the commercial kits—either the API 20E kit consisting of 23 tests (Part A) or the ENTEROTUBE II kit of 15 tests (Part B). The API 20E kit requires more manipulation and greater attention to aseptic technique than does the ENTEROTUBE II kit. Your instructor will determine which of these to use.

A. Multitest System: API 20E

Materials

1. API 20E kit, including[a]
    a. API 20E strip
    b. incubation tray and lid
    c. report sheet
    d. differential chart
2. 0.85% saline[a]
3. Sterile capped 13 mm test tubes
4. Sterile mineral oil[a]
5. Sterile applicator sticks
6. Sterile Pasteur pipets, 5 ml
7. Test tube racks
8. Marking pen
9. 50 ml plastic squeeze bottle with tap water
10. Zinc dust
11. H_2O_2, 1–5%
12. Kovac's indole reagent[a]
13. Nitrate A and B reagents[a]
14. Voges-Proskauer reagents[a]
15. Ferric chloride reagent[a]

16. Nutrient agar slant
17. Oxidase reagent
18. An unknown nutrient agar slant culture

[a] = Available from API

Procedure

Period 1

The API kit consists of microtubes containing dehydrated substrates. The addition of the bacterial suspension rehydrates these and the bacteria grow. Changes in the substrate can be determined by observation or the addition of reagents after 18–24 hours incubation at 35°–37°C. This kit can be used to determine some Gram-negative non-*Enterobacteriaceae* by modifying the incubation period and the addition of a few tests.

1. Bacterial suspension:
   a. Add 5 ml of sterile 0.85% saline to a sterile test tube.
   b. With a sterilized loop, transfer a small amount of growth to the saline and emulsify thoroughly. (In clinical practice this would be one well isolated colony from a selective or differential medium.)
2. Preparation of the strip:
   a. Set up an incubation tray and lid and put your name on the flap for identification.
   b. With the squeeze bottle, add 5 ml of tap water to the incubation tray to provide a humid environment during incubation (Figure 34-1).

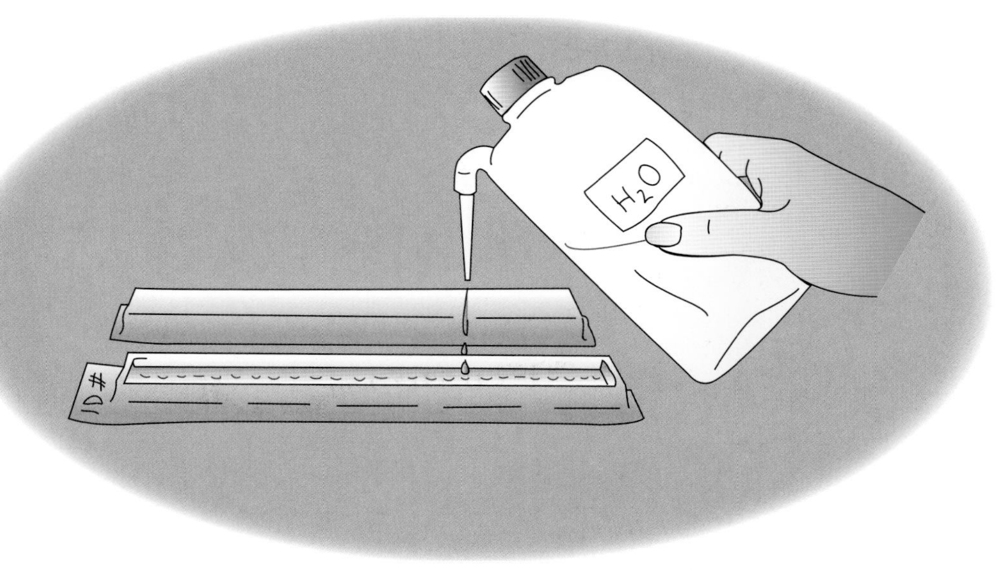

Figure 34-1

API tray

c. Remove the API strip from the sealed pouch and place it in the incubation tray (Figure 34-2).

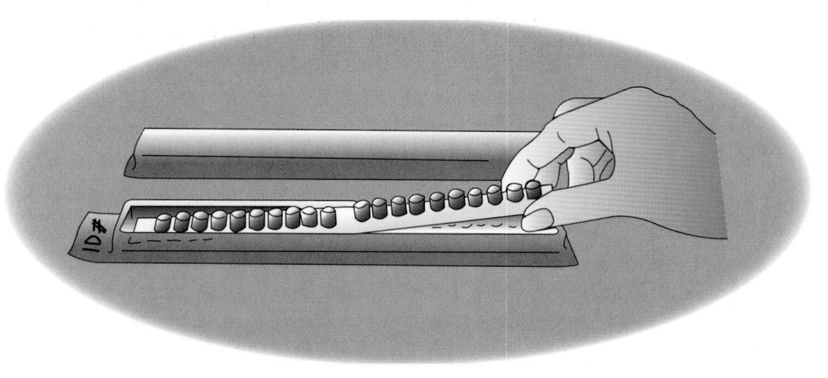

Figure 34-2

Adding API strip

3. Inoculation of the strip:
   a. The API 20E strip contains 20 microtubes consisting of a cupule and a tube (Figure 34-3, inset).

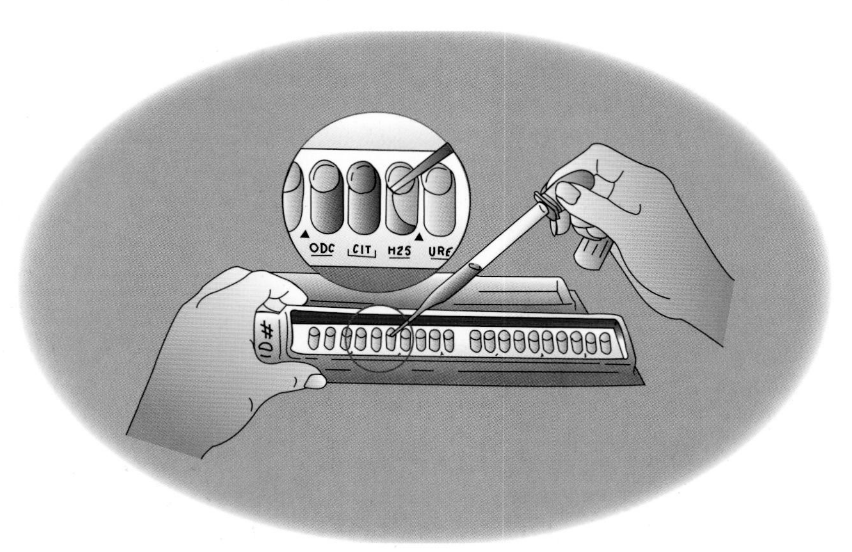

Figure 34-3

Tube and cupule detail; adding culture

   b. Using a sterile 5 ml Pasteur pipet with bulb, remove some bacterial suspension from the previously prepared tube.
   c. Tilt the incubation tray and fill the tube section of the microtubes by placing the pipet tip against the side of the cupule (Figure 34-3) **except** the *ADH, LDC, ODC,* H_2S, and *URE* microtubes, which are best interpreted by being

slightly underfilled. Fill both the tube and the cupule section of *CIT*, *VP*, and *GEL*.

d. After inoculation, fill the cupule section of the *ADH*, *LDC*, *ODC*, *H_2S*, and *URE* tubes with sterile mineral oil (Figure 34-4).

Figure 34-4

Adding mineral oil

4. With the remaining bacterial suspension, streak a nutrient agar slant. Incubate it with the strip.
5. Place the lid on the tray and incubate the strip for 18–24 hours at 35°–37°C. *Note:* strips may be removed from the incubator after 18–24 hours on weekends and stored at 28°C until Monday.

Observations (see Table 34-1)

Period 2

1. After 18 hours and before 24 hours of incubation, record all test reactions not requiring addition of reagents.
2. If the *GLU* tube is negative (blue or green), check with your instructor.
3. If the *GLU* tube is positive (yellow), note if gas bubbles are present and perform the oxidase test.
   a. Add a drop of oxidase reagent to a piece of filter paper.
   b. With a sterile applicator stick (do not use nichrome wire loop as it gives a false positive reaction), transfer some growth from the agar slant to the drop of oxidase reagent on the filter paper. A positive test is a purple color appearing within 30 seconds. If there is no change in color, the test is negative.

Table 34-1

Summary of Test Results, API 20E Strip (from API literature)

Tube	Positive	Negative	Comments
ONPG	yellow (any shade)	colorless	Orthonitrophenylgalactopyranoside
ADH	red or orange	yellow	Arginine dihydrolase
LDC	red or orange	yellow	Lysine decarboxylase
ODC	red or orange	yellow	Ornithine decarboxylase
CIT	turquoise or dark blue	light green or yellow	Citrate, read the cupule area (aerobic)
H_2S	black ppt	no black ppt	Hydrogen sulfide, browning is negative
URE	red or orange	yellow	Urea
TDA	brown red	yellow	Tryptophane deaminase. Add 1 drop 10% ferric chloride.
IND	red ring	yellow	Indole. Add 1 drop Kovac's reagent.
VP	red	colorless	Voges-Proskauer. Add 1 drop 40% KOH, 1 drop α-naphthol.
GEL	diffusion of pigment	no diffusion of pigment	Gelatin. Any diffusion is positive.
GLU MAN INO SOR RHA SAC MEL AMY ARA	yellow or gray yellow	blue or blue green blue or blue green	Fermentation occurs primarily at bottom, oxidation primarily at the top. GLU = glucose, MAN = mannitol, INO = inositol, SOR = sorbitol, RHA = rhamnose, SAC = sucrose, MEL = melibiose, AMY = amygdalin, ARA = arabinose.
GLU	note bubbles, add 2 drops nitrite A, 2 drops nitrite B		
nitrate	red	yellow	Add zinc
reduction Zn	yellow	red	
MAN INO SOR catalase	bubbles	no bubbles	Add H_2O_2 and observe for bubbles. Use tube with no gas and, if possible, no acid.

4. Add the reagents to the TDA and VP tubes (Figure 34-5). If positive, the TDA reaction will be immediate; whereas, the VP test may be delayed up to 10 minutes.
5. Add the Kovac's reagent to the IND tube last.
6. Perform the nitrate reduction test on all oxidase-positive organisms. This must be done last (see Table 34-1).
7. Record all results as positive or negative on the 24-hour line on the form accompanying the kit (or the Results and Observations form). The biochemical tests are grouped in

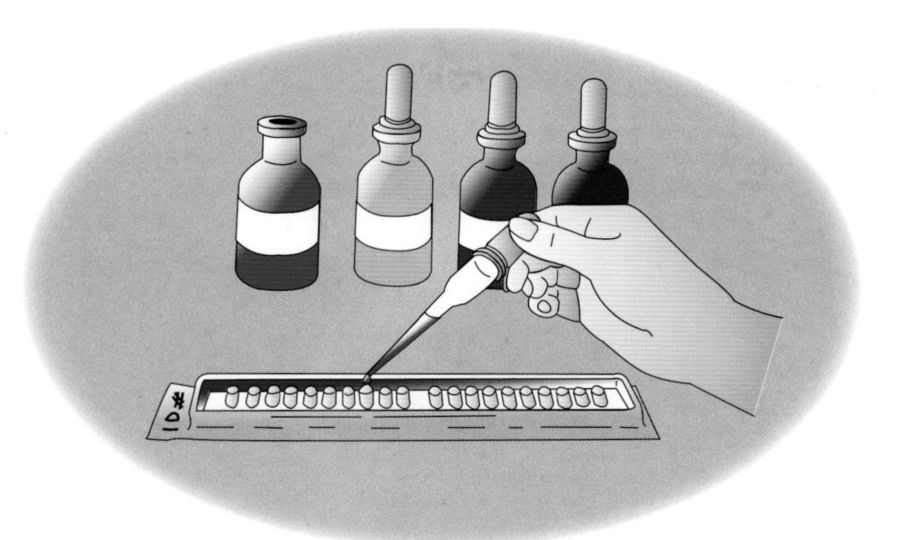

Figure 34-5

Reagent addition

threes and weighted, the numerical weights given as 1, 2, or 4 at the top of the column (see API form). Sum the weighted values (maximum 7) for all positive tests in the group of three and place the sum in the box on the line marked "Profile Number". Repeat until seven profile numbers have been generated. Compare this seven-number profile code with the Analytical Profile Index, Enterobacteriaceae supplied by API, or those provided on the results form and make an identification. Note: If oxidase-positive organisms are used, your instructor will supply you with additional media and instructions. Two additional digits can then be added to the Index code.

B. Multitest System: Enterotube II

Materials

1. 18- to 24-hour-old nutrient agar plate culture of the unknown organism
2. 1 ENTEROTUBE II unit
3. 1 code form (or use the code form on the results page)
4. ID Code Manual (one for class)
5. 3 plastic syringes and needles for indole and Voges-Proskauer reagents (for class)
6. Indole reagent (Kovac's)
7. Voges-Proskauer reagents:
    a. alpha-naphthol
    b. KOH solution

Procedure

Period 1

NOTE: These procedures are color-illustrated in the booklet accompanying the kit.

1. Remove both caps from the ENTEROTUBE II.
2. Touch the tip of the wire to the growth on the plate (Figure 34-6). Enough cells must be present to make a visible mass on the tip and side of the wire. Avoid touching the agar.

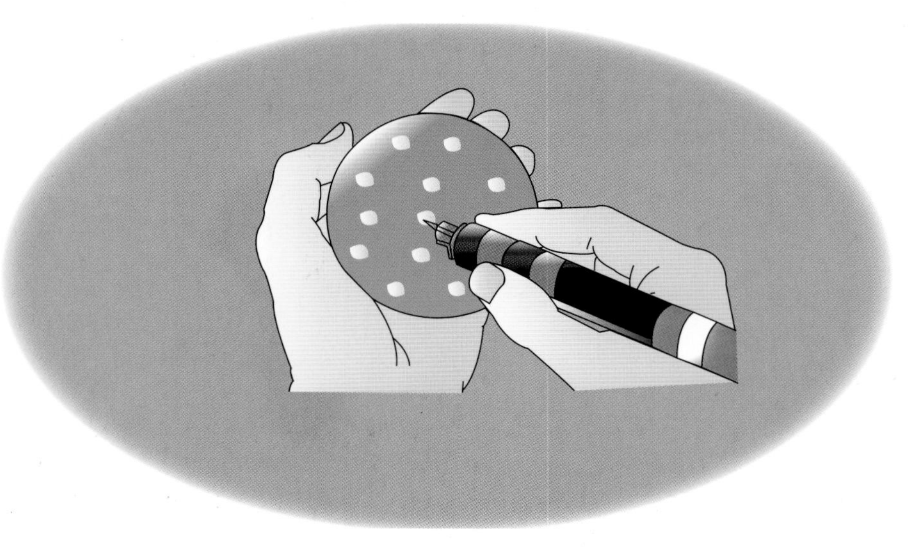

Figure 34-6

Picking a colony

3. Immediately withdraw the wire with a screwing or twisting motion through all 12 compartments (Figure 34-7).

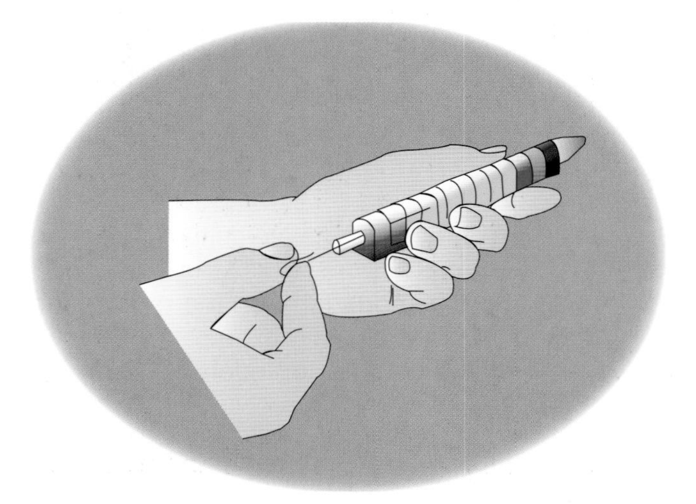

Figure 34-7

Inoculating

4. Reinsert the wire through all 12 compartments. Then withdraw the wire until the tip is in the H_2S/indole compartment. Now break the wire at the notch by bending. Discard the handle and replace the caps on the ends loosely. The wire remaining in the tube ensures anaerobic conditions and will not interfere with the reactions (Figure 34-8).

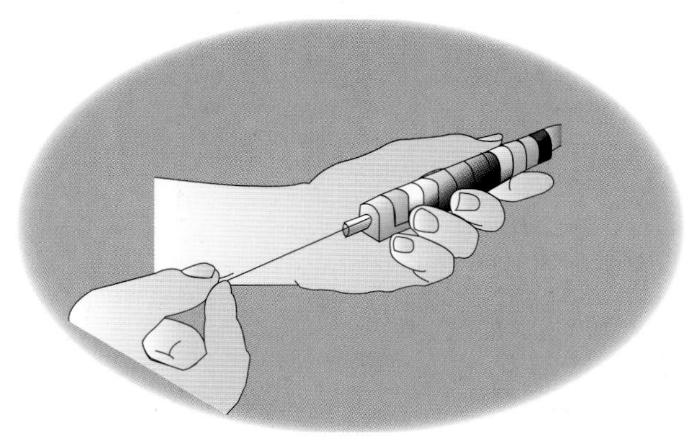

Figure 34-8

Reinserting and breaking wire

5. Strip off the blue tape after inoculation, but before incubation, to provide aerobic conditions in certain compartments. *Slide the clear band* over the glucose compartment to keep wax from escaping if gas is produced (Figure 34-9).

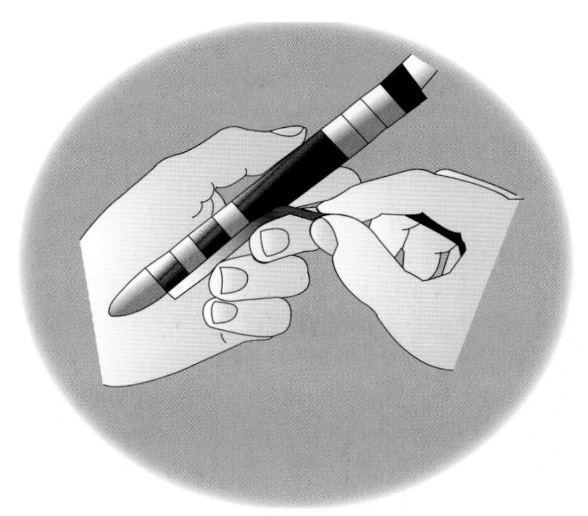

Figure 34-9

Replacing caps and removing blue strip

6. Incubate at 35°–37°C for 18–24 hours with the tube horizontal on its flat side.

Table 34-2

Summary of Test Results, ENTEROTUBE II Strip

Compartment	*Substrate*	*Positive Test*
GLU	glucose	any yellow color
GAS	gas	causes separation of wax, NOT by bubbles in the medium
LYS	lysine	decarboxylation causes a purple color
ORN	ornithine	decarboxylation causes a purple color
H_2S	thiosulfate	H_2S causes a black color
IND	tryptophan	reagent turns red in 10 seconds—MUST be last step
ADON	adonitol	fermentation causes yellow color
LAC	lactose	fermentation causes yellow color
ARAB	arabinose	fermentation causes yellow color
SORB	sorbitol	fermentation causes yellow color
VP	glucose	reagent turns red within 20 minutes if acetoin is present, orange is negative
DUL	dulcitol	fermentation causes yellow color
PA	phenylalanine	deaminase causes black to smoky gray color
UREA	urea	urease causes red-purple color
CIT	citrate	growth on citrate turns medium deep blue or light blue-green in some cases

Observations

Period 2

1. If available, use the color booklet accompanying the tubes and an uninoculated unit for comparison.
2. ***ALL* observations must be made before the indole and Voges-Proskauer tests are performed.**
3. After all other observations are made, perform the indole test by placing the tube in a rack with the glucose compartment down, then inject the indole reagent into compartment 4. If positive, it will turn red in about 10 seconds.
4. Finally inject the Voges-Proskauer reagents into compartment 9 with tube in a rack and the glucose compartment down.
5. Record the results on the ENTEROTUBE II form as + or – for each compartment on the form. If positive, circle the number immediately below the compartment.
6. In the hexagon below each arrow, enter the total points circled. If nothing was positive in a bracket, enter zero (0). The five numbers become the ID Value for your unknown.
7. Now go to the Computer Coding book and find the ID value matching your unknown. The name of the organism will be given along with any atypical results for your strain. If more than one organism is listed, a probability of identification will also be given. The confirmatory test column would provide further aid in identification.
8. Enter the information on the form and other pertinent data on the Reports and Observation form.

UNIT VII

Applied Microbiology Techniques

Putting microbes to use antedates microbiology as a science. The making of bread, beer, and wine are the oldest of the biotechnological uses of microbes. Newer uses involve production of proteins, enzymes, hormones, amino acids, vitamins, antibiotics, industrial chemicals, and many others. Here we look at several uses of microbes for the protection of public health and the food supply. The next unit on medical aspects is also an applied field but very specialized; for this reason it is treated separately.

Water covers 70% of the earth's surface; most of that is sea water with a salinity of about 3.5% chiefly NaCl. Although fresh water is a relatively small part of the total water volume, its importance to humans is out of proportion to its size because of its use for drinking water and for agricultural and industrial uses. One of the exercises in this section is a water quality test, widely used for determining the safety of water and its potential as a disease source.

It has often been said, "We are what we eat." In addition to being a source of nutrients, there are other important aspects of food. These include spoilage, foodborne illnesses, the number of organisms as a quality control factor, and the conversion of one food to another by encouraging the growth of certain microbes. This last category includes food fermentations, such as cheese, pickles, beer, wine, soy sauce, and salami, among others. Almost all foods contain living organisms, with the exception of canned foods and a few others. It is necessary that microbes be limited in number (except in fermented foods) lest the food spoil too soon and that the food be free of pathogens. Most people are surprised to learn that so many microbes are found in the food they eat. Several quality control exercises are presented in this unit as well as an example of a food fermentation.

35 Water Quality Analysis: MPN Method

Objectives

The student will be able to:

1. perform the MPN coliform test.
2. name the three steps in the MPN coliform test.
3. describe the media used in each step of the test.
4. describe a positive test for each step of the test.
5. define the term fecal coliform.
6. calculate an MPN from a seven-tube, three-dilution table.
7. recognize coliform growth and characteristic colonies.
8. name the alternate method used for drinking water.

Water can be analyzed for disease potential by performing the **MPN** (most probable number) test or the membrane filter test. These are quantitative bacteriological tests for a group of bacteria called **coliforms**, which occur in large numbers in the intestinal tract of humans and animals. While not normally pathogens themselves (a few are), they do indicate the presence of sewage or animal waste and thereby intestinal pathogens, because they come from the same site in the body. Almost any human and many animal intestinal pathogens, bacteria, viruses, and parasites can be transmitted through water, often leading to large epidemics. Among these are typhoid fever, cholera, salmonellosis, amebic dysentery, shigellosis, hepatitis, and others. The presence of pathogens in water is not common and often they are few in number so the coliforms are used instead to indicate the potential. The coliform group includes *Escherichia coli*, *Enterobacter aerogenes*, and a number of closely related species, all having the ability to ferment lactose with gas production.

The MPN test for coliforms consists of three steps: a presumptive test, a confirmed test, and a completed test (Figure 35-1). It attempts to determine the numbers of organisms in the water

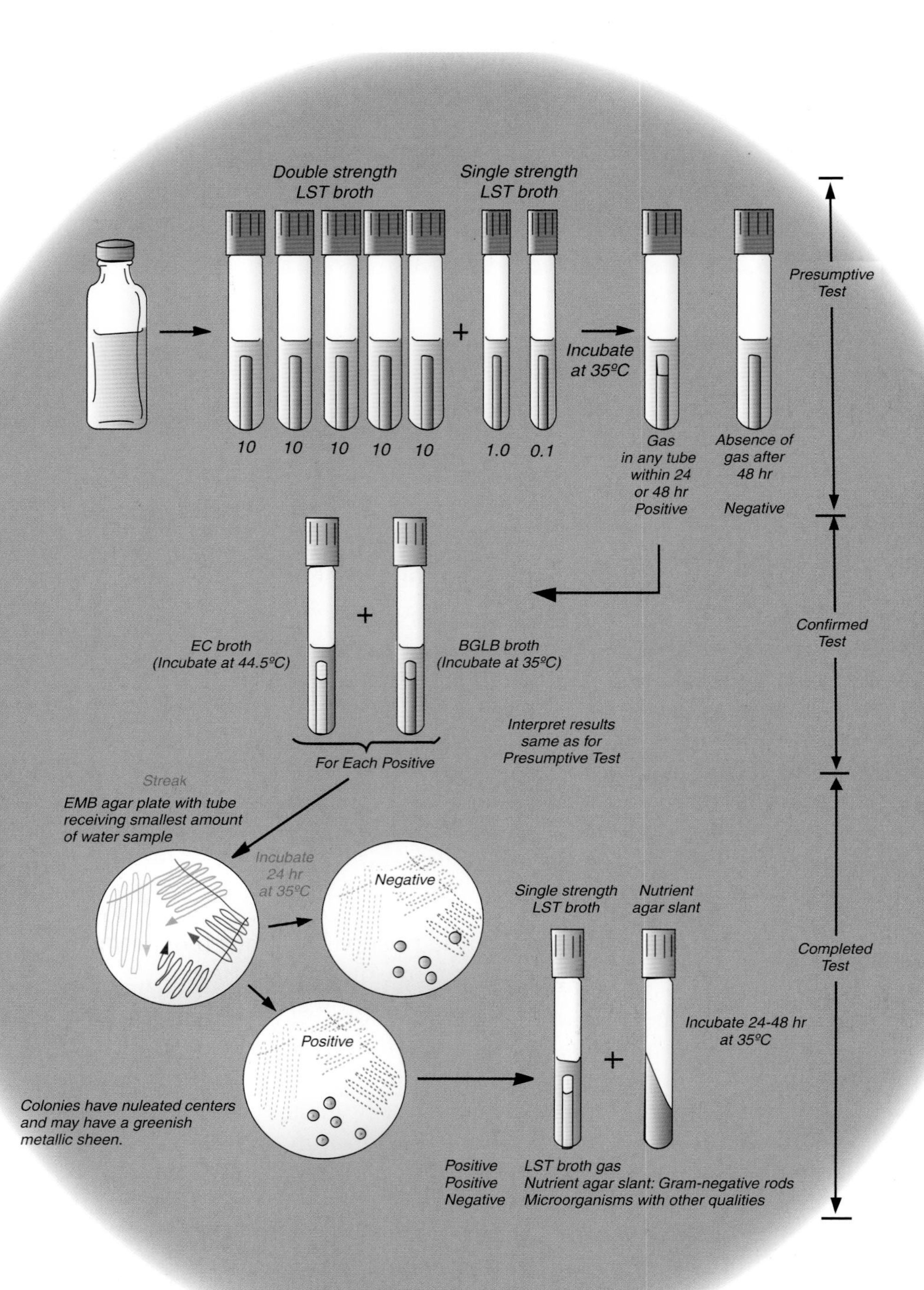

Figure 35-1

Outline of steps in the MPN coliform test

that are Gram-negative and ferment the carbohydrate lactose with the production of gas at 35°C. They must be facultative anaerobes and nonsporeformers.

The first step is the **presumptive test**. A set of tubes of lauryl sulfate tryptose lactose (LST) broth is inoculated with samples of water and incubated. Lauryl sulfate is a surface active detergent that inhibits the growth of Gram-positive organisms while encouraging the growth of coliforms. Coliforms use up any oxygen present in the broth and then ferment the lactose, producing acid and gas under anaerobic conditions. Gas formation in 24 or 48 hours is a positive test.

Positive tubes from the presumptive test are subcultured into brilliant green lactose bile (BGLB) broth to provide the **confirmed test**. BGLB broth, in addition to containing lactose, also contains two components inhibitory to Gram-positive bacteria. Brilliant green is a dye related to crystal violet and belongs to the triphenylmethane dye series. Ox bile is a surface active agent that also inhibits the growth of Gram-positive bacteria. Gas formation in 24 or 48 hours "confirms" the results of the presumptive step. The number of coliforms per 100 ml of water is then calculated from the distribution of positive and negative tubes in the test by referring to an appropriate table (Table 35-1). Results are reported as coliform MPN per 100 ml of water.

In some cases the organisms must be isolated and stained to provide the **completed test**. Positive BGLB tubes are streaked on eosin-methylene blue (EMB) agar. The two dyes, eosin and methylene blue, also inhibit the growth of Gram-positive organisms. Typical colonies (Table 35-2) are isolated on nutrient agar slants and inoculated into LST broth. If gas is now formed in 24 or 48 hours, a Gram stain is made from the growth on the slant. If the cells are Gram-negative and there is no indication of spores, the completed test is considered to be positive. Further biochemical studies (IMViC) may be performed on isolated cultures.

All three tests are necessary to prove that an organism in a water sample is in truth a coliform. In actual practice, when it has been shown that the presumptive and confirmed tests give essentially the same results, then the completed step is generally not done because of the time it takes.

A modification of the confirmed test allows enumeration of **fecal coliforms** (*Escherichia coli* Type I). These particular strains are closely associated with the human intestinal tract. Positive presumptive tubes are subcultured into EC broth in addition to the BGLB broth and incubated in a 44.5°C water bath. EC medium contains bile and lactose. The bile inhibits Gram-positive bacteria, while the high temperature selects only those organisms able to grow at this temperature. Gas in 24 hours is a positive test for fecal coliforms.

The official MPN method calls for 5 tubes at each of three

Table 35-1

MPN Index and 95 percent confidence limits

Number of Tubes Showing a Positive Reaction Out of			MPN Index	95 Percent Confidence Limits	
5 of 10 ml each	5 of 10 ml each	5 of 10 ml each	per 100 ml	Lower	Upper
0	0	0	<2	0	5.9
*0	0	1	2		
0	1	0	2	.05	13
*0	1	1	4		
1	0	0	2.2	.05	13
*1	0	1	4.4		
1	1	0	4.4	.52	14
*1	1	1	6.7		
2	0	0	5	.54	19
*2	0	1	7.5		
2	1	0	7.6	1.5	19
*2	1	1	10		
3	0	0	8.8	1.6	29
*3	0	1	12		
3	1	0	12	3.1	30
*3	1	1	12		
4	0	0	15	3.3	46
4	0	1	20	5.9	48
4	1	0	21	6.0	53
*4	1	1	27		
5	0	0	38	6.4	330
5	0	1	96	12	370
5	1	0	240	12	370
5	1	0	240	12	3700
5	1	1	>240		

*These tube combinations are unlikely. If they occur in more than 1% of the tests, it indicates faulty technique or other problem.

Source: *Standard Methods for the Examination of Water and Wastewater*, 13th ed., New York: The American Public Health Association, 1971.

Table 35-2

Differentiation of coliforms on EMB agar (see Figure 35-2)

	Escherichia coli	*Enterobacter aerogenes*
Size	Well-isolated colonies are 2–3 mm in diameter.	Well-isolated colonies are larger than *Escherichia coli,* usually 4–6 mm in diameter or more.
Confluence	Neighboring colonies show little tendency to run together.	Neighboring colonies run together quickly.
Elevation	Colonies slightly raised; surface flat or slightly concave, rarely convex.	Colonies considerably raised and markedly convex; occasionally the the center drops precipitously.
Appearance by Transmitted Light	Dark, almost black centers which extend more than 3/4 across the diameter of the colony; internal structure of central dark portion difficult to discern.	Centers deep brown; not as dark as *Escherichia coli* and smaller in proportion to the rest of the colony. Striated internal structure often observed in young colonies.
Appearance by Reflected Light	Colonies dark, button-like, Often concentrically ringed with a greenish metallic sheen.	Much lighter than *Escherichia coli,* metallic sheen not observed except occasionally in depressed center when such is present.

Source: Reprinted with permission from "Bacteria Fermenting Lactose and Their Significance in Water Analysis" by Max Levine. Iowa State College of Agriculture and Mechanical Arts Official Publication Vol. 20, No. 31, Bulletin 62, 1921.

dilutions for greater accuracy. However, in this exercise five tubes at the lowest dilution (largest sample volume) plus one tube at two others are used to conserve equipment.

Coliform organisms in **treated water** and many others also are generally determined now by the membrane filter method, which is quicker and employs a much larger volume of water. The MPN method still is the method of choice when water is turbid or contains many bacteria or algae; it is used also with foods, although even then a surface plate technique is now approved instead of the MPN method. There are other approved methods for certain sample sources (e.g., the fecal streptococcus group represented by *Enterococcus faecalis*).

This exercise will introduce the student to one of the most widely used methods of coliform determination in natural waters (it is also used with foods). The purpose of the membrane filter method is the same although the technique differs. Your instructor may elect to do only the Presumptive and Confirmed tests.

Materials

1. 1 empty sterile dilution blank. Collect 100 ml of a water sample from any stream, lake, pool, gutter, or other source. Samples should be collected on the day of the lab, or, if it is necessary, on the day before

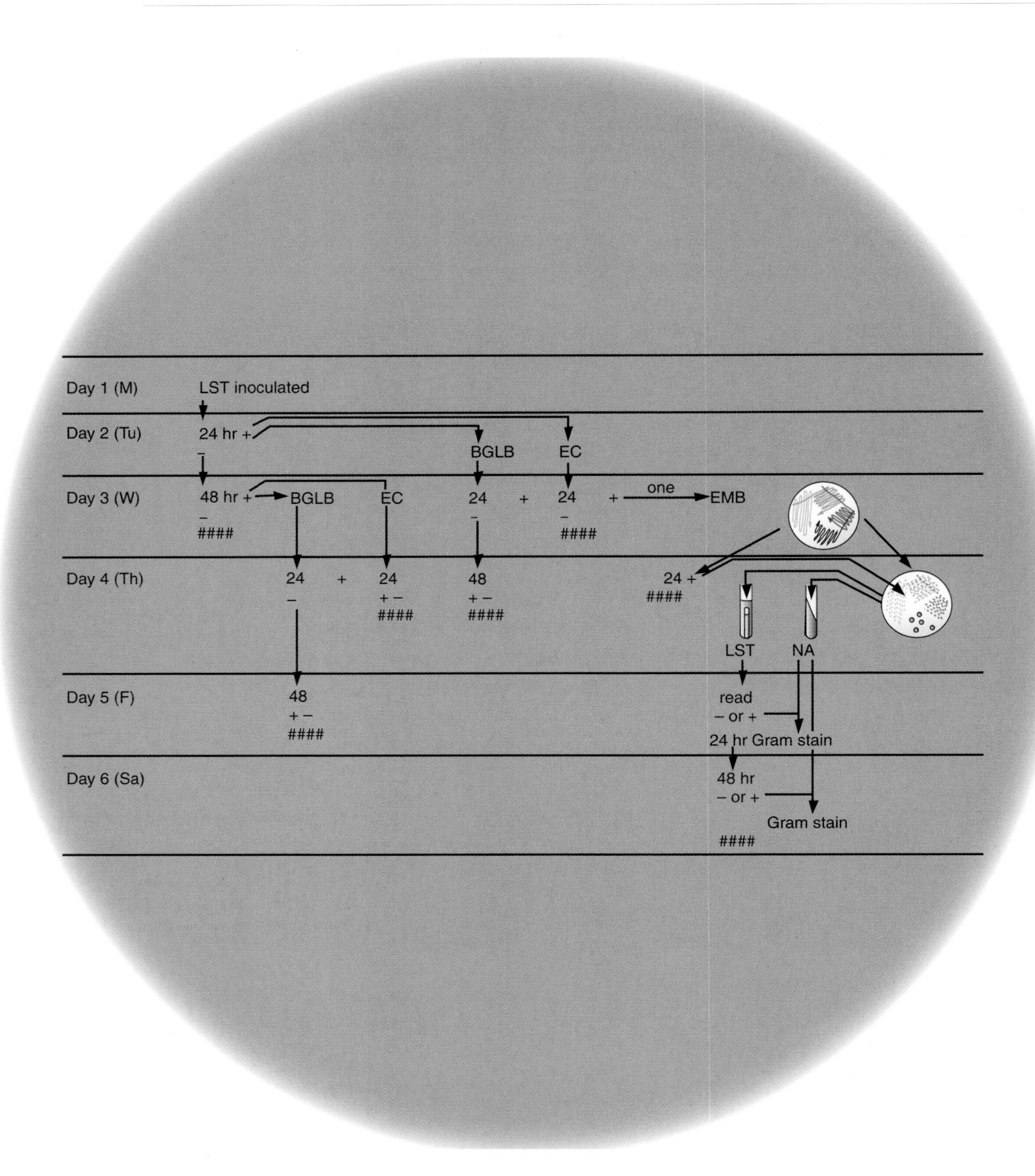

Figure 35-2

Coliform schedule and flow chart

and refrigerated until used. **Do not use drinking water or tap water** as it is chlorinated and is usually negative.

Presumptive and Confirmed tests

2. 5 tubes double-strength LST broth, sterile
3. 2 tubes single-strength LST broth, sterile
4. 1 sterile 10 ml pipet plus pipet bulb or aid
5. 1 sterile 1 ml pipet plus pipet bulb or aid
6. 7 tubes BGLB broth, sterile
7. 7 tubes EC broth, sterile

Completed test

8. 1 EMB agar plate, sterile
9. 2 tubes single-strength LST broth, sterile
10. 1 nutrient agar slant, sterile

Procedure and Observations

(See Figure 35-2 for daily schedule.)

A. Presumptive Test

Period 1

1. Shake the water sample 25 times in a one-foot arc in 7 seconds.
2. Pipet 10 ml into each of the 5 double-strength LST broth tubes.
3. Pipet 1 ml into one of the single-strength LST broth tubes and 0.1 ml into the second single-strength LST broth tube.
4. Incubate the tubes at 35°C.
5. At the end of 24 hours, agitate each tube vigorously to release any gas dissolved in the medium. The presence of any gas is a positive presumptive test. Record your results in the table provided. Negative tubes must be reincubated for an additional 24 hours.

B. Confirmed Test and Fecal Coliforms

Periods 2–4

1. For each LST broth tube that shows gas at 24 or 48 hours, transfer one loopful of growth to a BGLB broth tube and a second loopful to an EC broth tube.
2. Incubate the BGLB broth tubes at 35°C and the EC broth tubes in a thermostatically controlled water bath at 44.5°C.
3. The EC broth tubes are read at 24 hours. Gas is a positive test. Record your results in the table provided.
4. Read the BGLB broth tubes at 24 and 48 hours. Gas is a positive test. Record your results in the table provided. Save one or more positive tubes for Part C.
5. Determine the MPN for total coliforms using the distribution of positive and negative tubes in the confirmed test and the MPN table provided (Table 35-1).
6. Determine the MPN for fecal coliforms using the distribution of positive and negative tubes in the fecal coliforms test and the MPN table provided (Table 35-1).

C. Completed Test

Periods 3–6

1. In practice, this step would be done for each positive BGLB tube. Only one tube will be used here.
2. Streak one positive (gas) BGLB tube (or EC tube) for isolation on an EMB agar plate. Incubate at 35°C for 24 hours.
3. A typical isolated colony (see Table 35-2 for description) is selected and inoculated into one tube of single-strength LST broth and streaked on one nutrient agar slant. These are incubated at 35°C and observed at 24 and 48 hours.
4. If no gas appears in 48 hours, the test is negative. If gas appears in the LST broth tube insert, a Gram stain is made of growth on the nutrient agar slant culture. If cells are Gram-negative and no spores are present, the completed test is positive for that tube. If spores are present, the culture must be re-isolated on EMB; proceed again as in Step 3.
5. Record your results in the table provided.

D. IMViC Reactions (Optional)

1. If desired, some idea of the species of coliforms may be gained by applying the IMViC tests to the nutrient agar slant culture.
2. At least two more re-isolations on EMB should be done to insure purity before proceeding.
3. Refer to Exercise 23 for procedures of these tests.

36 Standard Plate Count of Food

Objectives

The student will be able to:

1. prepare, dilute, and plate a food sample.
2. count colonies according to the rules for counting plates.
3. explain the results to a friend.
4. discuss in a general way the reasons for concern with bacteria in food.

Bacteria and fungi occur in most foods in varying quantities. Generally, numbers are fairly low—less than 100,000 per g or ml depending on the food—unless growth has occurred, such as in fermented or spoiling food. Spoiling food may contain 10^7 or more per g or ml. The number of bacteria—or **colony forming units (CFU)**—is determined by homogenizing a food sample in a diluent, making a dilution series (Exercise 10), and plating the dilutions with **standard plate count agar (PCA)**. The plates are then incubated, and the number of bacteria (or CFU) are determined per g or ml. An evaluation can then be made of the quality of the food based on the normal condition of that food. **Large** numbers (10^5–10^7) can generally be interpreted to mean that the food has been held under conditions **permitting growth** of bacteria. In many foods, especially previously cooked foods, this suggests that foodborne illness organisms may also have grown. However, a food may be **dangerous** to health with very little growth, and the standard plate count may not be very useful in such cases. In many situations, selective media for particular pathogens can be used instead. Standards for some foods have been established using the plate count (e.g., milk). The standard plate count is not useful with all foods, such as fermented foods (cheese, pickles, etc.) in which organisms are encouraged to grow, reaching 10^7 or more per g or ml. Again, special media for specific groups can be used in these cases.

Two food types will be studied in this exercise: hamburger

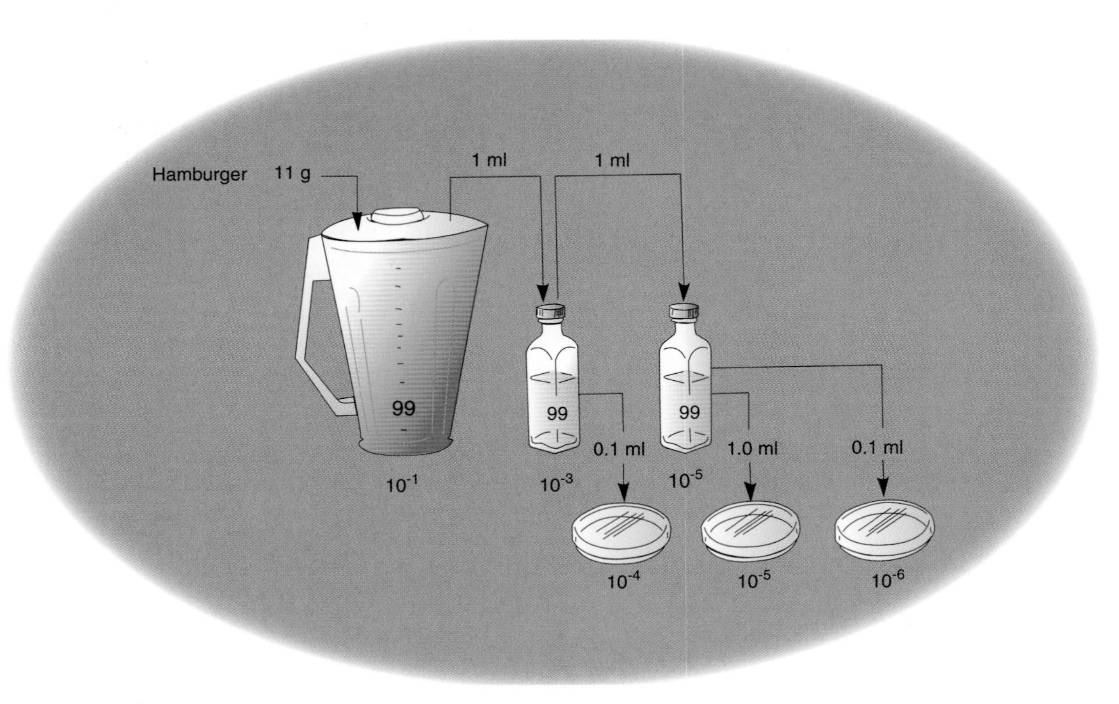

Figure 36-1

Hamburger dilution scheme

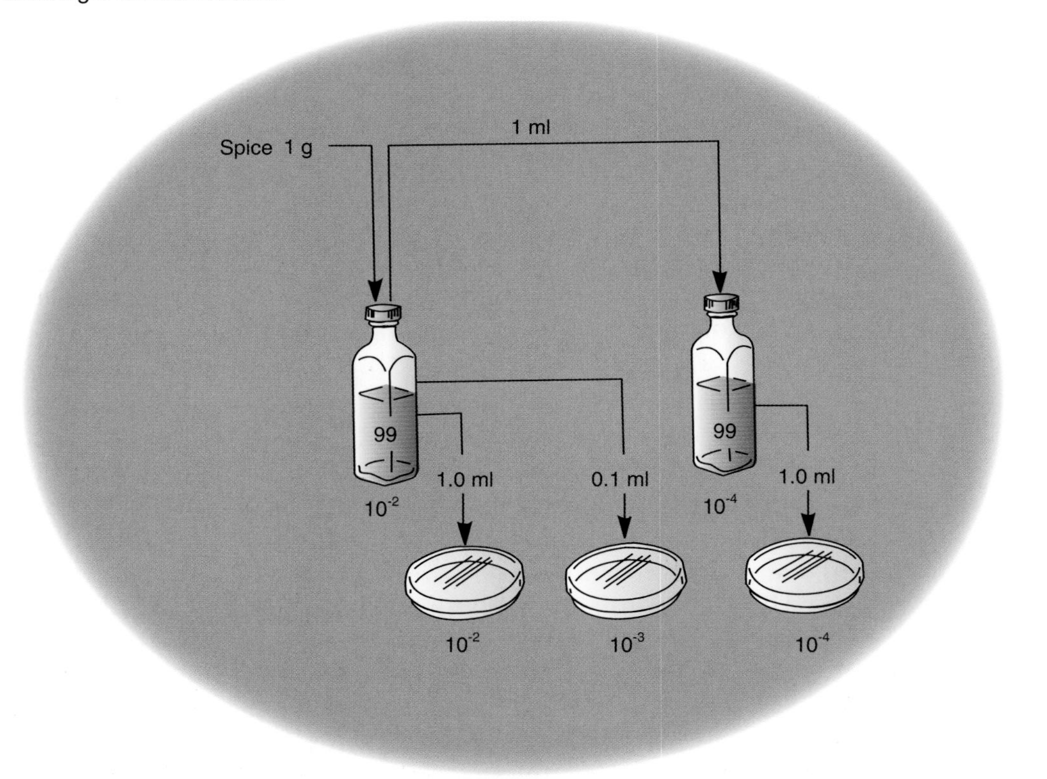

Figure 36-2

Spice dilution scheme

and two spices, black pepper and chili powder. The results will provide the student with some information on the bacteria in commonly used food items. The use of spices in many cultures evolved as a means of masking off-odors and flavors; they have little effect in commonly used amounts on the growth of organisms, food poisoning, or otherwise. A few spices (cinnamon and cloves) contain antibacterial agents, but they are used in such small amounts as to be ineffective in the food.

Materials (groups of 3 pairs)

1. A sample of fresh hamburger from a local market and cans of black pepper and chili powder
2. 1 Waring blender jar with aluminum foil cover, sterile (for hamburger)
3. 7 sterile 99 ml dilution blanks (hamburger 3, pepper 2, chili 2)
4. 9 sterile Petri plates
5. 7 sterile 1-ml pipets with aid
6. 1 Flask plate count agar (PCA) for 9 plates, sterile
7. Spatulas

Procedure

(Review Exercise 10 for dilutions and counting procedures)
Period 1

Hamburger (Figure 36-1)

1. Weigh 11 g of hamburger aseptically into a sterile blender jar with a flamed and cooled spatula.
2. Add one 99 ml dilution blank to the jar and blend 1–2 minutes at low speed and 1 minute at high speed. This is the 10^{-1} dilution.
3. Pipet 1 ml of the 10^{-1} dilution to a second 99 ml dilution blank labeled 10^{-3}. Discard the pipet and shake the blank—25 times in a one-foot arc in 7 seconds. Pipetting the mixture may prove difficult. Let the mixture settle, then tip the dilution blank to the side and use the supernatant.
4. Pipet 1 ml of the 10^{-3} dilution to a third blank making the 10^{-5} dilution. Discard the pipet and shake the blank as described.
5. With a new pipet, transfer 0.1 ml of the 10^{-5} dilution to a plate labeled 10^{-6}. Then with the same pipet transfer 1 ml of the 10^{-5} dilution to a plate labeled 10^{-5}, blowing out the last drop. With the same pipet transfer 0.1 ml of the 10^{-3} dilution to a plate labeled 10^{-4}. Discard the pipet.
6. Pour enough tempered (45°–50°C) PCA into each plate to just cover the bottom (about 20 ml). Mix carefully as instructed in Exercise 10 and let solidify.
7. Incubate the plates at 32°C for 48 hours (or 5 days at room temperature).

Black Pepper and Chili (Figure 36-2)

1. Weigh 1 g of spice directly into a 99 ml dilution blank with a flamed and cooled spatula. Shake vigorously 25 times in a one-foot arc in 7 seconds. This is the 10^{-2} dilution.
2. Pipet 1 ml of the 10^{-2} dilution into a second 99 ml dilution blank labeled 10^{-4}. Discard the pipet and shake the blank vigorously as described.
3. Pipet 1 ml of the 10^{-4} dilution into a plate labeled 10^{-4}.
4. With the same pipet, transfer 0.1 ml of the 10^{-2} blank to a plate labeled 10^{-3}.
5. With the same pipet, transfer 1 ml from the 10^{-2} blank to a plate labeled 10^{-2}.
6. Pour enough tempered (45°–50°C) PCA into each plate to just cover the bottom (about 20 ml). Mix carefully as instructed in Exercise 10 and let solidify.
7. Incubate the plates at 32°C for 48 hours (or 5 days at room temperature).

Observations

Period 2

1. Using the rules for counting plates in Exercise 10, count all the colonies on qualified plates using a colony counter and record the results on the record form.
2. Note any differences between colony morphology on the plates for the three foods.
3. Make a Gram stain of cells from a few surface colonies from each food and record the Gram reaction and morphology of each.

37
Standard Plate Count of Milk

Objectives

The student will be able to:

1. identify five general sources of milk contamination.
2. list 5 bacterial diseases transmitted to humans through milk.
3. discuss the methods of milk pasteurization.
4. perform a standard plate count of milk.
5. calculate the plate count per ml of milk.

Milk contains a variety of nutrients and has a pH of 6.8, thus it is a good medium for the growth of many different bacteria including pathogens. Milk has a small **initial microbial flora** at the time it is drawn from the cow. However, unless good sanitation practices are followed, milk may be contaminated by many more microorganisms, especially those that cause disease. Such **sources of contamination** include the cow, the milking area, milking equipment, personnel, and processing.

In order to make milk safe for human consumption, **pasteurization** is applied to commercially processed milk. There are three methods of pasteurization. The oldest and least used today is the **holding or vat method**, which consists of raising the temperature of milk to 63°C and holding it there for 30 minutes. The most commonly used method today is **flash pasteurization**, which involves the heating of milk to 72°C and holding it for 15 seconds. A third method is **ultra-high-temperature pasteurization,** using temperatures above 121°C for 1–2 seconds. Regardless of the method employed, the pasteurization procedure is designed to kill pathogenic microorganisms found in milk—particularly *Coxiella burnetii,* the causative agent of Q fever and the most heat resistant of these milkborne pathogens. Note that pasteurization does not kill all bacteria. Many **diseases** are potentially transmitted to humans in milk: **from the cow**—tuberculosis (bovine), brucellosis, *Streptococcus, Salmonella,* campylobacte-

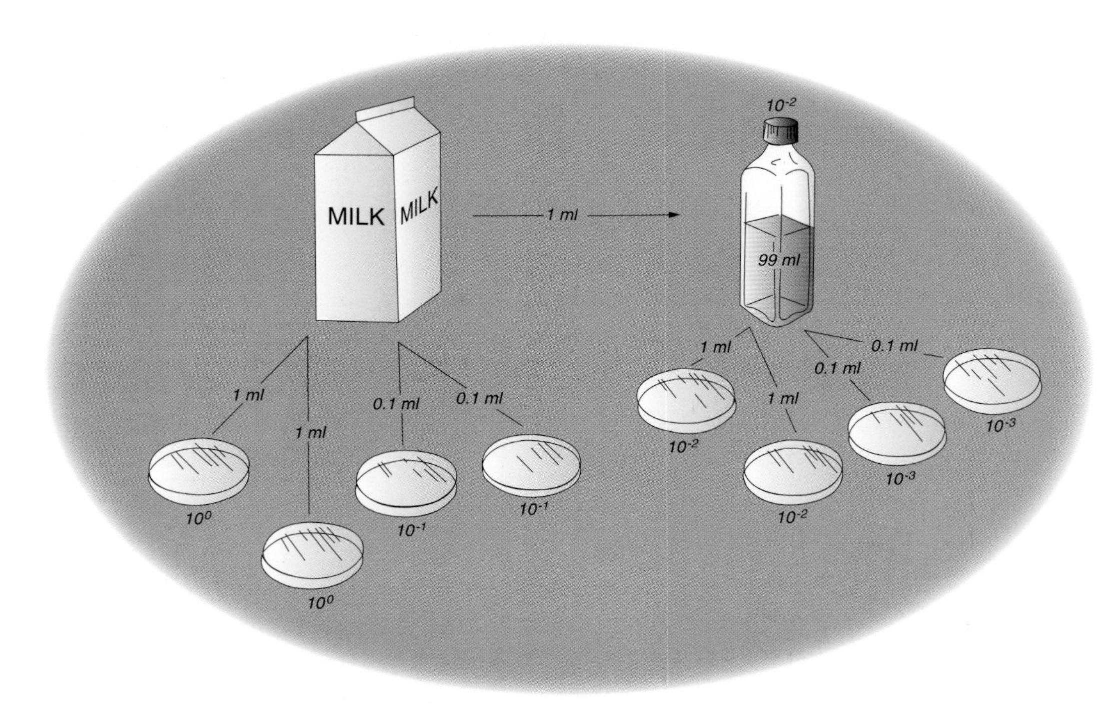

Figure 37-1

Preparation of dilution plates for pasteurized or raw milk

riosis; **from human milk handlers**—typhoid fever, scarlet fever, diphtheria, tuberculosis (human), hepatitis; **from soil and dirty equipment**—listeriosis, *Salmonella*, and campylobacteriosis, to name a few. Some of these will grow in the milk; others need only be present to cause illness.

The quality and safety of milk are monitored by state and local health departments using standards set forth by the U.S. Food and Drug Administration, the American Public Health Association, and the dairy industry. One of the widely used tests is to determine the total number of bacteria in a ml of raw or pasteurized milk as an indication of how the milk was handled at the farm, during shipment, and after pasteurization. This test is known as the **standard plate count** and uses **standard plate count agar** (PCA) tested under specific conditions and approved for this use. According to accepted standards, raw milk (from individual suppliers) may not have more than **75,000 bacteria per ml** before pasteurization and must have **less than 15,000 per ml** after pasteurization. Note that a well-run milk farm today will produce raw milk with a total count of **less than 10,000 per ml**. Just because a raw milk has a low bacterial count **does not** mean it is safe. Public health records show that pathogens may be present even if the total count is 1,000–2,000 per ml, which is the level found in most modern dairies.

The purpose of this exercise is to introduce the student to a widely used method for monitoring the quality of milk. The procedure can be used for both pasteurized and raw milk. In most cases, raw milk will not be available in the local market area due to health restrictions on its sale, so pasteurized milk is the sample of choice. Select a market milk that is as fresh as possible, since the total number of bacteria rises with time while refrigerated.

Materials (per pair)

1. Carton of pasteurized milk or a sample of raw milk
2. 1 sterile 99 ml dilution blank
3. 8 sterile Petri plates
4. Flask of plate count agar, sterile
5. 2 sterile 1 ml pipets and aid

Procedure (2 days)

Period 1

1. Label eight Petri plates in pairs for each of the following dilutions: 10^0, 10^{-1}, 10^{-2}, and 10^{-3}. Include your name, date, etc., as well (Figure 37-1).
2. Before opening, shake the container of milk 25 times in a one-foot arc in 7 seconds.
3. Aseptically transfer 1 ml from the carton of milk into a 99 ml dilution blank and then discard the pipet. This transfer results in a 10^{-2} dilution. Shake the blank as before.
4. Pipet 1 ml from the carton of milk into each of the 2 plates labeled 10^0. Do not discard the pipet.
5. Using the same pipet from step No. 4, transfer 0.1 ml from the carton of milk into each of the 2 plates labeled 10^{-1}. Discard the pipet.
6. Select another pipet and transfer 1 ml from the 99 ml dilution blank into each of the 2 plates labeled 10^{-2}. Do not discard the pipet.
7. Using the same pipet from step No. 6, transfer 0.1 ml into each of 2 plates labeled 10^{-3}. Discard the pipet.
8. Introduce 15–20 ml of cooled plate count agar into each plate. As each plate is poured, thoroughly mix the medium with the aliquot of milk sample by rotating the plates gently in a figure-8.
9. After solidification, invert the plates and incubate at 32°C for 48 +/– 3 hours.

Observations

Period 2

1. After incubation, count the plates with colony numbers between 30 and 300, using the procedure and rules of Exercise 10.
2. Enter the counts on the report form and calculate the standard plate count (SPC) per ml of milk.

38
Preparation and Analysis of Yogurt

Objectives

The student will be able to:

1. make yogurt.
2. make a quantitative viable plate count of the number of organisms present in the product.
3. make a quantitative direct microscopic count of the organisms present in the product.
4. identify by name and describe the morphological appearance of cells and colonies of the two bacteria involved.

Yogurt and acidophilus-fermented milk products have been used for centuries by peoples of the Near East. Very likely it was originally a means of preserving milk, which incidentally (and unknown to the users, of course) killed disease organisms as a result of the lactic acid produced.

Yogurt fermentation is brought about by inoculation of boiled milk with two lactic acid bacteria, *Streptococcus thermophilus* and *Lactobacillus bulgaricus* (Figure 38-1). The final product is quite acidic, with a white appearance and a custard-like texture due to coagulation of milk protein by lactic acid. Fruit or flavors mask the acidity and improve acceptability.

Acidophilus fermentation is brought about by inoculation of boiled milk with *Lactobacillus acidophilus.* The final product is white, extremely acid (much more than yogurt), and somewhat granular or lumpy. Fruit or flavors generally improve acceptability.

Foods and many other substances are often examined directly for the presence of bacteria. It is useful to see and count organisms that may be dead and unable to grow (e.g., a heated food involved in a foodborne illness outbreak). **Direct microscopic** examination is a technique that allows one to stain and count organisms present in the sample. A major limitation of this technique is the need for at least 500,000 per gram of food in order to see enough cells under the microscope to make a reliable count. This limits its use to foods in which bacteria have grown to sizable numbers.

This exercise is intended to introduce the student to the making of yogurt and to analysis of the product for bacterial content, using the standard plate count and direct microscopic count methods.

Materials

Preparation of Yogurt

1. Powdered skim milk and tablespoon
2. 1 liter beaker or flask
3. Milk supplied by student (500 ml or 1 pint)
4. Long-handled teaspoon or dowel for stirring
5. 2 plastic 200 ml (6–8 oz.) hot cups or glass jars with covers
6. Plain yogurt inoculum (grocery store)
7. Thermometer (non-mercury)

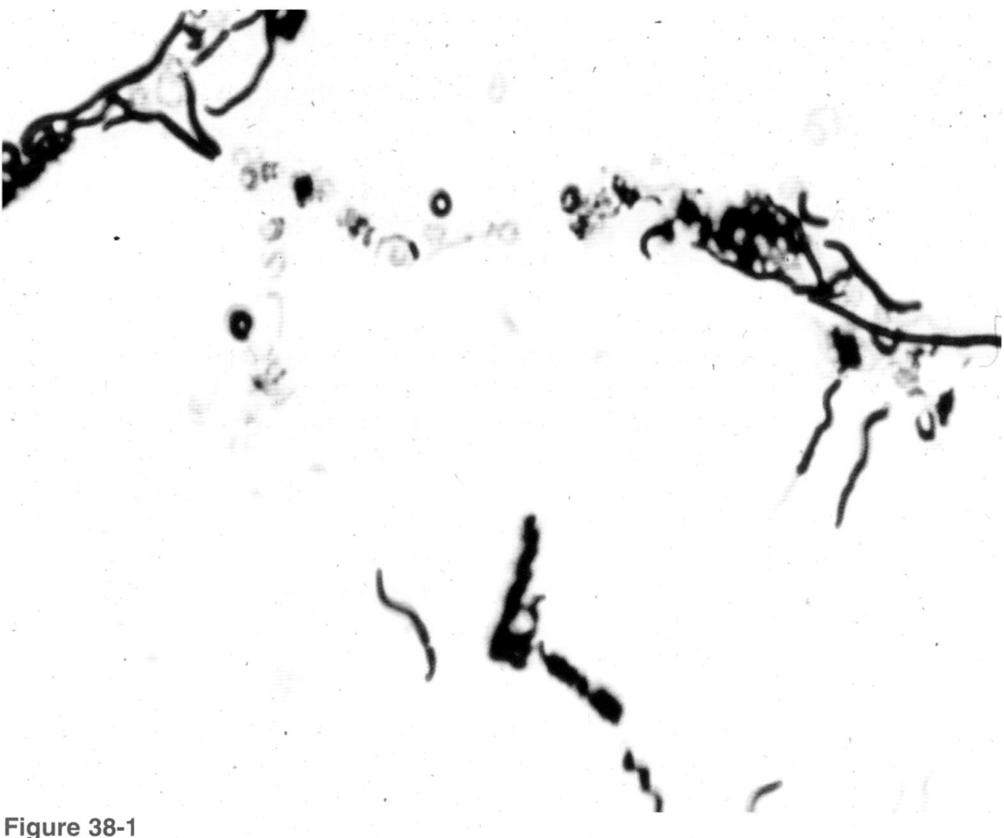

Figure 38-1

Lactobacillus bulgaricus on LAB agar.

Analysis of Yogurt

1. 1 sterile 0.1 ml pipet and aid
2. 2 sterile 1 ml pipets
3. 4 sterile 10 ml pipets
4. 5 sterile 99 ml dilution blanks
5. Spatula
6. 3 sterile Petri plates
7. LAB agar flask, sterile
8. Bent glass rod
9. Candle jar
10. 3% H_2O_2
11. Thermometer

Procedure (3-5 days)

Making of Yogurt or Acidophilus Milk

Period 1

1. Obtain 500 ml (one pint) of fresh, pasteurized milk (whole or skim). Larger volumes may be made by increasing the proportion of each item correspondingly.
2. Add at least 2 level tablespoons of powdered skim milk. More can be added if a firmer product is desired.
3. Bring the mixture to a boil over medium heat for 30 seconds, stirring constantly. Add a thermometer and cool to 45°–46°C (112°–115°F).
4. The inoculum used in class is a carton of grocery store plain yogurt without added flavor. Use about 1–2 teaspoonfuls of the store product per 500 ml of milk. Use a spoon boiled in the original milk or separately.
5. Mix well with the boiled spoon.
6. Pour the mixture into preboiled warm cups or dishes and cover with aluminum foil or other closure. Plastic hot cups with covers can be used directly. Fill to within several millimeters of the rim. Incubate at 42°C (109°F) for 9–15 hours or until desired firmness is obtained. If a 42°C incubator is unavailable in the lab, use the 37°C incubator and incubate for 18 hours. At home, use an oven at 42°C with a pilot light or an electric bulb or an electric oven with bulb. Change the light bulb wattage until the desired temperature is obtained. Use a thermometer to check the temperature in any case.

Period 2

7. After incubation, place the cups in a refrigerator until cool and serve. Sugar, flavors, jam, or preserves may be added, if desired. One cup should be saved without flavor for Part II below.
8. Samples of the above batch (without added flavors) may serve as an inoculum for the next batch and be stored in the refrigerator up to 10 days. Use 1–3 teaspoonfuls (boiled spoon) of this per liter

Figure 38-2

A candle jar

(quart) of newly boiled milk for continued culturing. An inoculum carried this way gradually deteriorates, producing a less desirable product. Replenish the inoculum periodically from a new commercial source.

Analysis of Yogurt

Period 2

A. Viable Count

1. After refrigeration for 1–2 days, prepare a 10^{-1} dilution. Weigh the yogurt container and remove 11 g with a flamed, cooled spatula, adding it aseptically to a 99 ml dilution blank. Shake well (see Exercise 10) 25 times in a one-foot arc in 7 seconds. Save this and subsequent dilution blanks for the direct microscopic count below.
2. Make the 10^{-3} through 10^{-6} dilutions in steps of 10^{-1}.
3. Pour three plates of LAB agar and let solidify.
4. Pipet 0.1 ml from each of the 10^{-6}, 10^{-5}, and 10^{-4} dilutions onto the surface of appropriately labeled LAB agar plates and spread with a sterile bent glass rod.
5. Incubate the plates at 37°–42°C for 2–4 days in candle jars (Figure 38-2) or a CO_2 incubator, if available.

B. Direct Microscopic Count

1. With a marking pen and a template, outline a one-centimeter-square area on a clean glass slide.
2. Pipet 0.01 ml of the 10^{-1} dilution onto the square cm area. With a bent inoculating needle (NOT a loop), spread the drop evenly over the area. Dry the film in a warm place.
3. Fix and stain the dried smear by the Gram method and examine the slides under oil immersion.

Observations

Period 2

A. Description of the Product

1. Make organoleptic observations of odor, texture, and taste, and record these on the report form.

B. Direct Microscopic Count

1. Make the following observations on 10–20 fields (if there are too many bacteria to count, prepare another slide from the next higher dilution):
   a. Count the total number of single bacterial cells and clumps of bacteria in each field and determine an average per field.
   b. Describe the various morphological types of bacteria in each field, count them, and average.
   c. The average count per field can be converted to count per g of original sample using the direct microscopic formula:

(Average count/field) x (dilution factor) x (microscope factor) = count/g

To determine the microscope factor, use a micrometer disk and measure the diameter (d) of the microscope field in millimeters (mm). Calculate the area: $mm^2 = \pi(d/2)^2$. Then calculate the microscope factor (MF) using the following formula where 100 = no. of 0.01 ml volumes/1 ml and 100 = no. of mm^2/1 cm^2.

$$\frac{100 \times 100}{mm^2 \text{ field area}} = MF$$

2. Enter data from the microscopic examination in the table provided.

Period 3

C. Viable Count

1. After incubation of the plates in the candle jars, make the following observations:

a. Count colonies on the plates and describe the dominant types. Large, opaque white colonies are *Streptococcus thermophilus*; small, translucent colonies are *Lactobacillus bulgaricus*.
b. Examine representative colonies by making Gram stains. Describe the morphology of organisms found.
c. After completing the above, flood the plate with 3% H_2O_2 and count colonies as catalase-positive (showing gas formation) and catalase-negative (no gas formation).

2. Enter data from the plate counts in the table provided.

39
Replica Plating for a Nutritional Mutant

Objectives

The student will be able to:

1. define auxotroph.
2. explain the purpose of mutating a bacterium to produce a desired auxotroph.
3. perform an experiment to produce an auxotroph requiring an amino acid.

Wild type organisms are those that are isolated from nature and grow on a complete or minimal medium. Mutations in the parent wild type (**prototroph**) often occur in nutritional pathways, and the progeny of these cells are unable to grow on the complete medium without supplementing with the newly required nutrient. These mutations often cause changes in the control mechanisms for metabolic pathways, not only leading to new nutrient requirements but also may result in increased production of some end product of a pathway. These nutritional mutants are called **auxotrophs** (*auxo* [G] food, *troph* [G] nourishment). Auxotrophs have been extensively used to study metabolic pathways, enzyme structure and function and are extensively exploited in biotechnology to produce large quantities of metabolic products.

Simple selection of strains for a desired trait or suppression of an undesired one from among those spontaneous mutants occurring in cultures has been used to obtain modified microbes. Deliberate genetic mutation for loss of one or more characters is done using ultraviolet light or chemical mutagens as the modifying agents.

This exercise illustrates the principle of **replica plating** and the use of a potent chemical mutagen to produce a nutritional mutant. The chemical **N-methyl-N'-nitrosoguanidine** (**MNNG**) generally causes mutations by the replacement of GC bases with AT bases in DNA. The reverse change may sometimes occur, as well as transitions, transversions, and frameshifts at low frequencies. **CAUTION: MNNG is also a carcinogen and should be handled with great care.** Using the replica plating described below, thousands of colonies can be tested for auxotrophic mutations (i.e., mutants able to grow on a complete medium but not on a medium deficient in some nutrient). MNNG may yield up to 20% total auxotrophs or 1–2% of a particular nutrient-requiring mutant.

The replica plating technique allows the transfer of a large number of colonies simultaneously to many different media to test nutritional requirements. A piece of sterile velveteen is attached to a wooden block the shape and size of a Petri plate bottom (Figure 39-1). The velveteen is placed on the surface of a master plate with many colonies; the fibers act as inoculators. The block is then transferred to another plate and touched to the surface, transferring some cells in the process. It is then removed and touched to a another plate, and so on. Each plate contains a different assortment of nutrients. Because both the block and the plate are marked, a plate showing no growth for a particular colony site can be traced back to the master plate and the colony there isolated for further study.

The bacterium used in this exercise (*Escherichia coli* K12 ATCC 25404) is capable of growing on minimal salts medium with amino acids. You will attempt to isolate an auxotrophic mutant requiring one of several amino acids to grow.

Materials (per pair)

1. 1 peptone broth centrifuge tube, sterile
2. 2 peptone broth tubes, sterile
3. 1 tube tris-maleic acid buffer, pH 6.0, sterile
4. 1 pair latex or rubber gloves
5. Solution of N-methyl-N'-nitrosoguanidine (MNNG) 1 mg/L, sterile
6. 7 sterile 1 ml pipets and safety aid
7. 2 sterile 99 ml phosphate buffer dilution blanks
8. 3 sterile 9 ml phosphate buffer dilution blanks
9. 1 glass spreading rod
10. 1 flask peptone agar, sterile
11. 8 Petri plates, sterile
12. 2 pre-dried plates each of minimal salts agar plus the amino acid combinations in Part II.1 below (10 plates), sterile
13. Replica plating block with rubber band or other restraining device
14. Sterile velveteen (a commercially prepared block and velveteen may be supplied instead)
15. Indelible marking pen with narrow tip
16. 18- to 24-hour-old peptone broth culture of *E. coli* K12

Procedure (about 6 days)

I. Mutagenesis

(Your instructor may elect to supply you with already mutagenized cells. In this case begin with step 10 below.)

Period 1

1. Inoculate a centrifuge tube containing 10 ml of peptone broth with the *E. coli* culture.
2. Incubate at 35°–37°C on a rotary shaker to the logarithmic stage of growth, about 12–15 hours (absorbance will be about 0.3). The cell count will be about 5 x 10^8 cells/ml.

Period 2

3. Centrifuge the culture for 5 minutes at 5000 x g and remove the supernatant aseptically.
4. Resuspend the cells by aseptically pouring in 9 ml of tris-maleic acid buffer at pH 6.0.
5. **CAUTION: Put on protective latex or rubber gloves**. Add 1 ml of freshly prepared MNNG, giving a final concentration of 100 µg/L. Incubate at 35°–37°C for 30 minutes without shaking.
6. Centrifuge the cells as before. **NOTE: Dispose of the MNNG supernatant in the special container provided for that purpose. It will be disposed of later as a hazardous waste.**
7. Resuspend the cells by aseptically pouring in 10 ml of peptone broth into the tube.
8. Centrifuge again. **NOTE: Dispose of the supernatant in the special container as in step 6 above.** Resuspend the cells by aseptically pouring 10 ml of peptone broth into the tube.
9. Incubate the medium and cells at 35°–37°C for 24–48 hours to allow the mutagenized cells to grow and express any mutations present.

Period 3

10. Pour 8 plates of peptone agar and let them dry thoroughly, inverted, overnight at 35°–37°C.
11. Prepare a serial dilution of the incubated culture from the last step in period 2 to include the 10^{-5}, 10^{-6}, 10^{-7}, and 10^{-8} dilutions as illustrated in Figure 39-2.
12. With a sterile 1 ml pipet, pipet 0.1 ml of the 10^{-8} dilution onto the surface of each of the labeled, pre-dried, duplicate peptone agar plates.
13. Repeat with the 10^{-7}, 10^{-6}, and 10^{-5} dilutions onto separate duplicate plates.
14. Spread the volume on the plates with an alcohol-flamed bent glass rod, beginning with the highest-dilution plates. Allow the liquid to absorb into the medium.
15. Incubate the plates inverted at 35°–37°C for about 24 hours until the colonies are >1 mm in diameter.

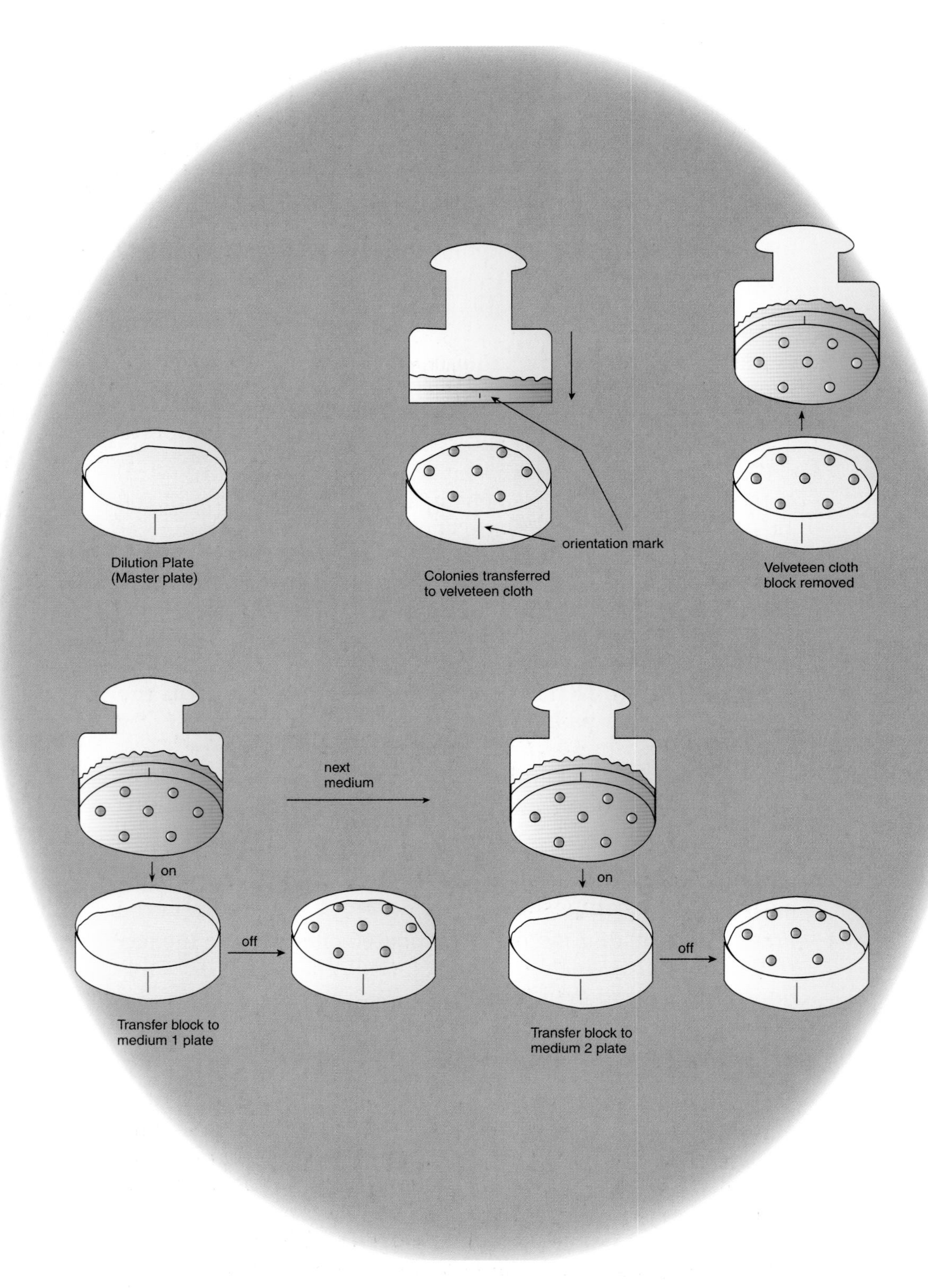

Figure 39-1

Replica plating procedure

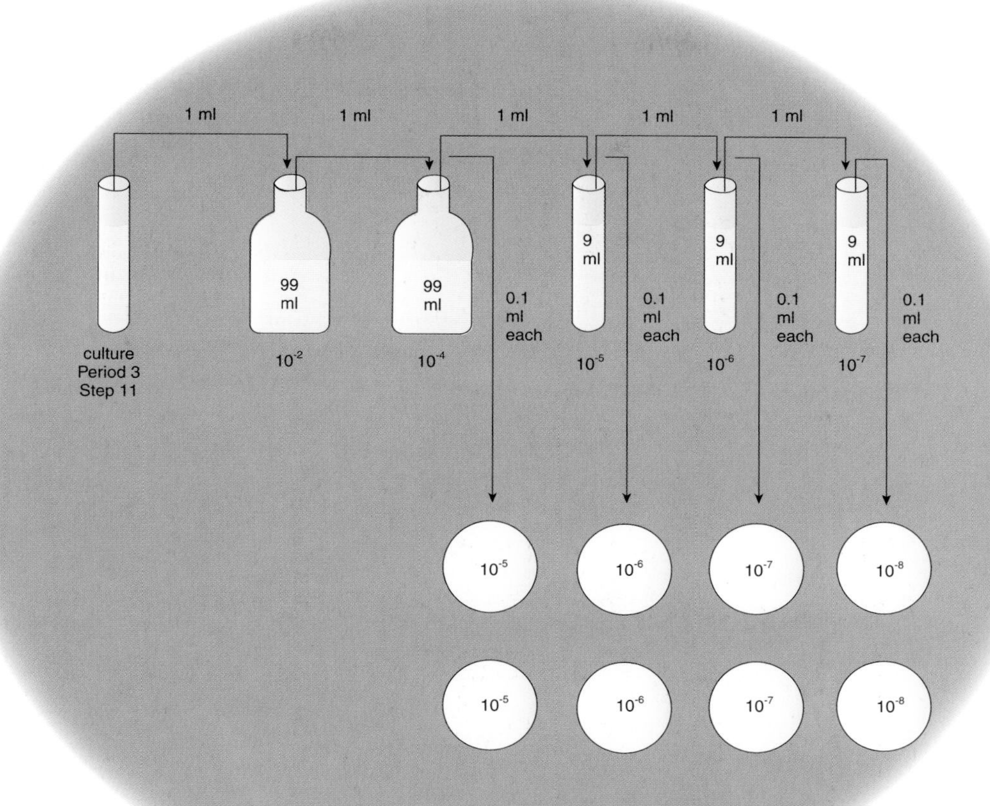

Figure 39-2

Dilution sequence for replica plating

II. Replica Plating

Period 4

1. Five minimal salts agar plates will be provided with the following amino acid compositions:

No.	*Composition*
1.	thr + met + leu + ala
2	met + thr
3	leu + ala
4	met + leu
5	thr + ala

2. Unless a commercial unit is supplied, prepare the replica plating block by placing a square of velveteen on the block and stretching it taut. Be careful not to touch the face of the cloth with your fingers. Hold the cloth in place with a rubber band or other retaining device.
3. Make a mark on the **edge** of the velveteen with an indelible pen (Figure 39-1). This mark will serve to orient the block.
4. Make a corresponding mark on the side of the bottom of each Petri plate (Figure 39-1). Next to the mark, put the number of

the medium from the table above.

5. Replica plate one of the dilution plates containing the closest to 20–50 colonies onto the five amino acid plates according to Figure 39-1. Be sure the orientation marks are lined up. This ensures that each plate will be oriented exactly the same way.
6. Incubate the five plates at 35°–37°C until colonies appear.

Observations

Period 5

1. Plate 1 serves as the Master plate. Make a drawing of this plate on the results form showing the location of **each** colony. Do the same for each remaining plate, drawing each colony. If a colony did not grow on a medium, use a light dashed line around the spot where it should have been. Make sure the orientation marks are included in each drawing.
2. Using Table 39-1, interpret growth on the plates.

Table 39-1

Growth of auxotrophs of a specific type on media with amino acid pairs

Type Colony	2 met+ thr+	3 leu+ ala+	4 met+ leu+	5 thr+ ala+
met⁻	+	–	+	–
thr⁻	+	–	–	+
leu⁻	–	+	+	–
ala⁻	–	+	–	+

Interpretation: A met⁻ auxotroph will grow on media 2 and 4 but not 3 and 5. A thr⁻ auxotroph will grow on media 2 and 5 but not 3 and 4; etc.

III. Isolation and Verification

Period 6

3. Select a colony from plate 1 that does not show growth on one of the amino acid plates. Streak the colony for isolation on plates of medium number 1 and the four amino acid media.
4. Incubate the plates at 35°–37°C for 24 hours or until colonies are well developed. Growth should occur on medium number 1 and not on at least one other.

Notes: It may be necessary to examine all the plates of the class to find one auxotroph. In that case, the colony (colonies) can be shared with the class for this part. Sometimes revertants occur that may be seen as a few isolated colonies on plates where the Master plate streak was not expected to grow. Keep an eye out for these.

Period 7

5. Record your results on the report form.

40 Bacterial Transformation

Objectives

The student will be able to:

1. describe the meaning of vector, competent cell, cloning, gene, plasmid amplification, recombinant, and genetic engineering.
2. clone a plasmid having an antibiotic resistance factor.

Plasmids are double-stranded, closed, circular DNA molecules occurring extrachromosomally in a variety of bacteria. These genetic elements carry a number of phenotypic characters often advantageous to the host organism, such as resistance to antibiotics and heavy metals, degradation of polymers and other complex organic molecules, and production of toxins, colicins, and antibiotics. Transfer of plasmids to new hosts can be accomplished in nature by a process similar to **conjugation** and in the laboratory by **transformation** to **competent** host cells (i.e., cells that are temporarily permeable and able to receive the plasmid). In nature, this may confer a selective advantage to the recipient cell and its progeny.

Plasmids are now widely used as **cloning vectors** for production of proteins. A vector is created by ligating a gene onto an enzymatically opened plasmid DNA strand and closing it again, a process called **splicing**. The **recombinant** plasmid is then inserted into a competent host cell and the host cell replicated, thus **amplifying** the vector. The amplification of the plasmid and the gene it contains is called cloning of the gene.

This exercise will demonstrate the cloning of a plasmid and the above concepts. No new genes will be spliced into the plasmid (a complicated procedure), but the method of cloning the plasmid is the same as used in modern **genetic engineering**. The plasmid used in this exercise has a gene-specifying resistance to the antibiotic ampicillin (related to penicillin). Since reagent preparation is

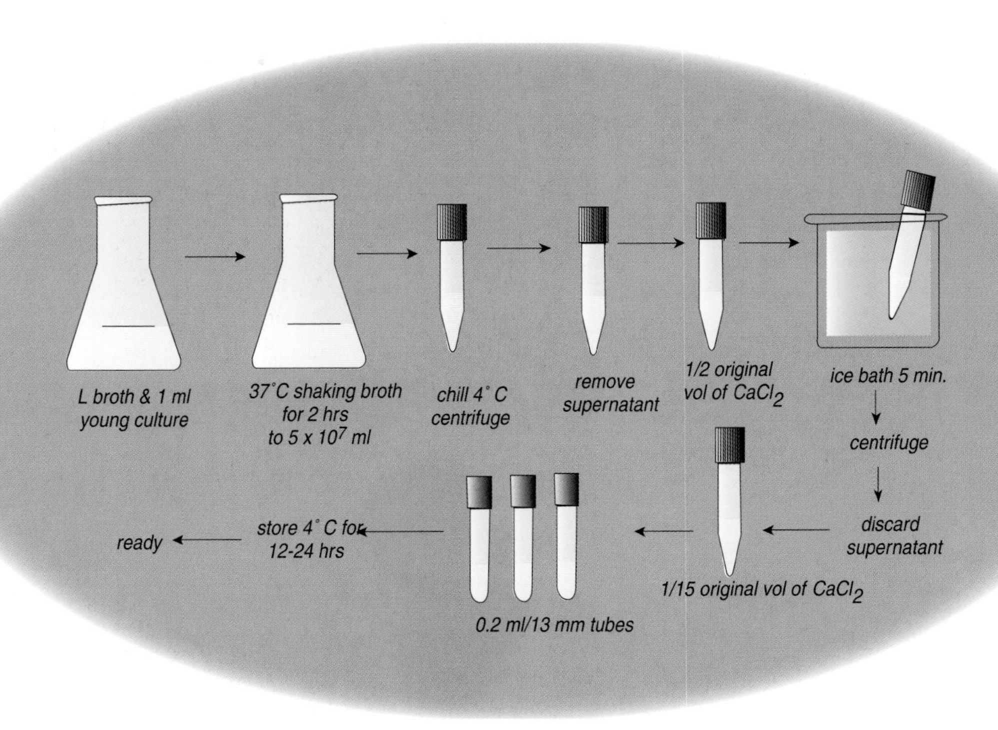

Figure 40-1

Preparation of competent host cells

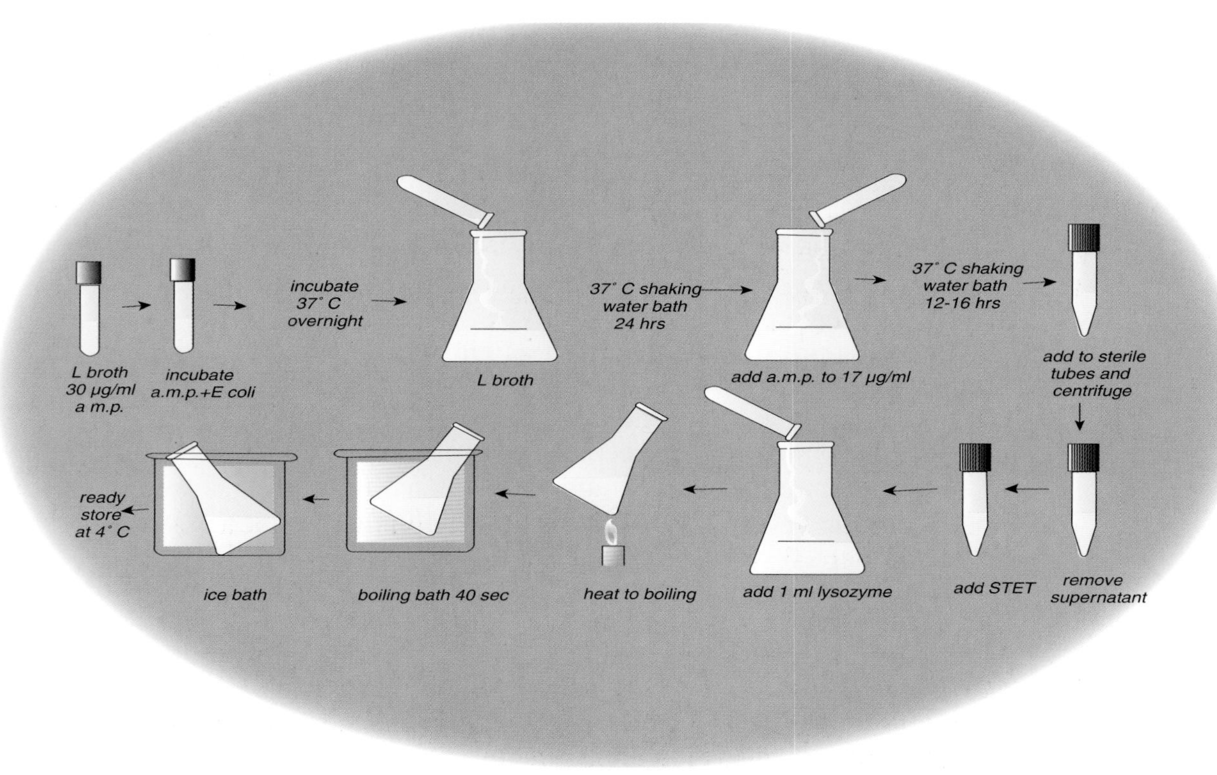

Figure 40-2

Plasmid amplification

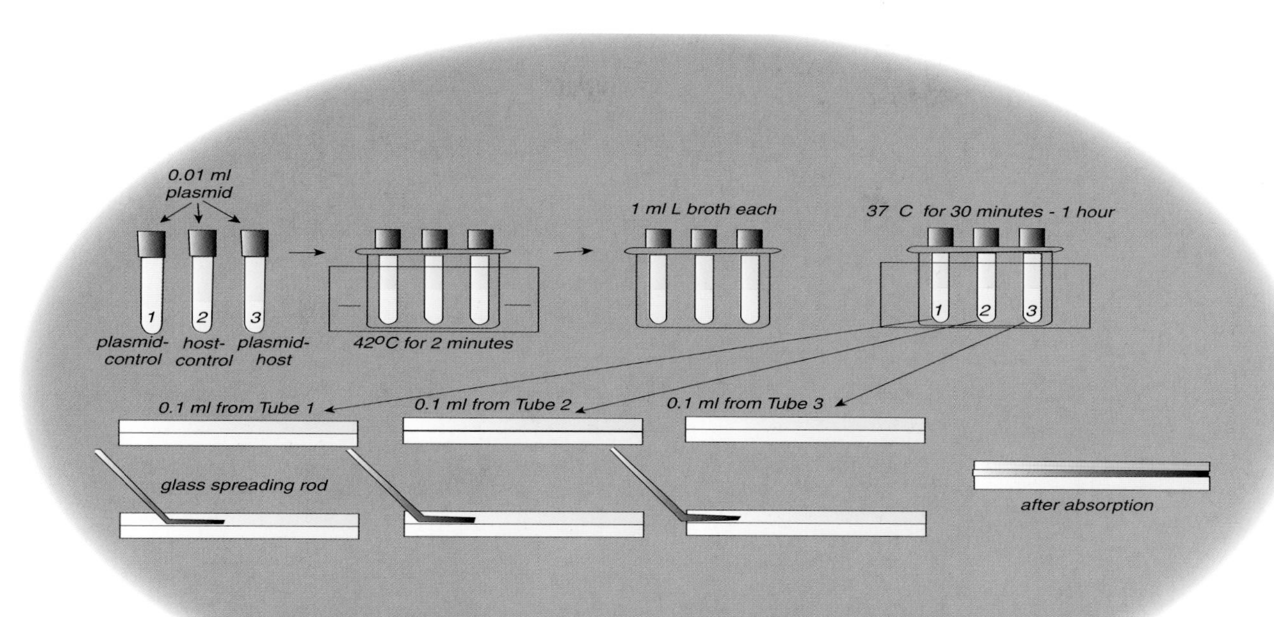

Figure 40-3

Flow diagram for transformation of *E. coli*

time consuming and expensive, the exercise makes use of a commercial kit that is reliable and easy to use. So that you will have some idea of the preparation involved, Figures 40-1, 40-2, and 40-3 have been included.

Materials (per six students, pairs, or by demonstration)

This exercise is based entirely on a kit prepared by Carolina Biological Supply Company. All reagents, equipment, and pipets, are included as part of the kit.

Procedure

Period 1
The procedure will not be described in detail here but will be found as part of the commercial kit described above.

Observations

Period 2
Count colonies on each plate and record your results on the report form using the format suggested in the kit.

UNIT VIII

Medical Microbiology and Immunology

One of the great successes of microbiology has been in the understanding and conquest of infectious diseases. Although there are still many diseases, particularly viral ones, where no cures have been found, many can be prevented and infectious disease is not the threat to life that it once was. This has resulted from success in identifying a particular organism as the cause of a specific disease, understanding how the organism is transmitted, and discovering antibiotics for treatment of disease and chemicals to kill or inhibit microbes.

The human body, as with all parts of the natural environment, is a habitat for numerous microbes, some just waiting for an opportunity to cause trouble. The teeth, throat, hands, and intestine are prime sources of problems intimately involving microbes. The use of antibiotics, antiseptics, and germicides to control bacteria on or in the body and elsewhere requires knowledge of their effectiveness and mode of action. The emergence of new disease agents and the development of resistance to chemical agents and antibiotics requires continual effort to stay ahead of the problems.

An integral part of understanding how the body fights off infectious agents is found in immunology, the study of the immune system. Louis Pasteur was an early pioneer in this fascinating area. Although it is necessary for the microbiologist to have some understanding of immunology, it is now a discipline of its own. Microbiologists are often called upon to perform blood tests of various kinds related to immune function. Some of the immune tests are widely used in identifying microbes or their products. It is essential that some knowledge of this important area be included in a microbiology laboratory manual.

Antibodies produced by B-cells in the blood react with specific **antigens**. This reaction can be observed in the laboratory using special serological (serum) techniques. **Agglutination** results when antibodies and large particles combine to form a large clump. **Precipitation** results when antibody combines with a small antigen forming a fine precipitate.

This unit provides some experience in analyzing the microbial flora of various parts of the human body, some tests on chemical and antibiotic sensitivity of selected microbes, and two widely used immunological procedures involving antibodies and antigens.

41 Normal Flora of the Human Throat

Objectives

The student will be able to:

1. properly perform the method of obtaining a throat culture specimen and subsequent isolation of microorganisms from a throat swab.
2. list four significant cultural characteristics that can serve as an aid in the identification of the isolates.
3. list at least four organisms that are part of the normal flora of the human throat.
4. name the differential medium used in the cultivation of a throat specimen.

Under normal conditions the mouth and throat contain **considerable numbers** of microorganisms. Most of these are harmless, but some are potential pathogens if given the right opportunity. Sometimes virulent organisms (e.g., *Streptococcus pyogenes*, which often causes "strep throat" and produces beta hemolysis) may be present but not producing disease, although the same species may be the cause of disease in other individuals.

Organisms commonly isolated from healthy individuals include *Staphylococcus* spp., *Streptococcus* spp., *Proteus* spp., diphtheroid bacilli, *Moraxella catarrhalis*, *Klebsiella pneumoniae*, other Gram-negative bacilli, lactobacilli, and *Haemophilus influenzae*, among others. The **viridans** group streptococci, which produce alpha hemolysis, are the most prominent of the throat bacteria, and a few of these are able to cause disease if given the right conditions. The numerically most prominent organism in the throat is often not the cause of a disease condition. In other words, the disease may be caused by an organism present in relatively small numbers.

Because most of these organisms have fastidious growth requirements, a **differential** medium (e.g., blood agar) is used for their cultivation. The hemolytic reaction along with Gram reaction,

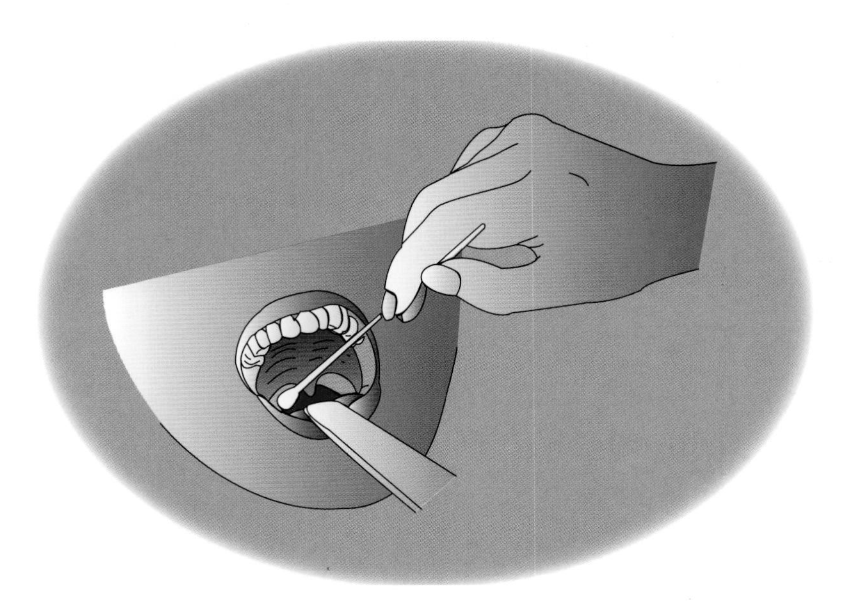

Figure 41-1

Proper method for taking a throat culture

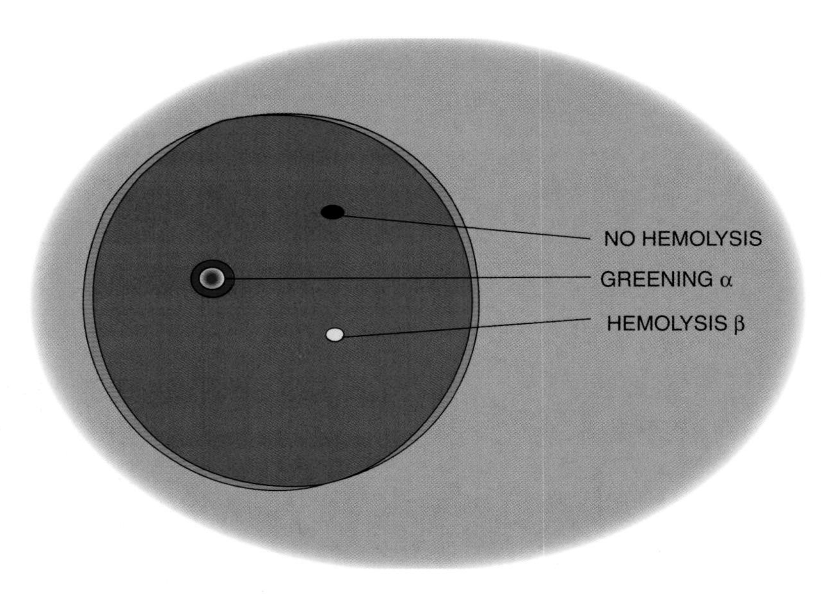

Figure 41-2

Types of hemolysis

colony size, and color given by a bacterial species growing on a blood agar plate can be used as a valuable aid in the identification of certain pathogens. In practice, swabs may be taken and transported to the laboratory in a tube of **transport medium** to prevent dehydration. See Exercise 20 for additional material on these hemolytic blood reactions.

Materials (pairs)

1. 2 sterile 5% sheep blood agar plates
2. 2 sterile cotton swabs
3. 2 tongue depressors
4. Clean microscope slides
5. Gram stain reagents

Procedure

Period 1

1. Divide the blood agar plates into 2 segments.
2. Have your partner swab your throat with a sterile cotton swab. Use a tongue depressor to hold the tongue down and to assure easy access to the throat. Insert the swab through the mouth to the tonsillar area without touching the tongue or any other oral surface. You should constantly rotate the swab as you obtain the specimen (Figure 41-1).
3. Take the swab with your throat specimen, streak and rotate over the first sector (original inoculum sector) of the blood agar plate. Discard the swab.
4. Using a sterile inoculating loop, streak the original inoculum sector over the remaining sector so as to obtain well-isolated colonies.
5. Incubate the plate at 37°C for 24 hours.

Observations

Period 2

1. After incubation, examine the plate for both alpha- and beta-hemolytic colonies (Figure 41-2). Make smears and Gram stain any hemolytic colonies of either type. Examine the stained slides carefully and note Gram reaction, cell shape, and cell arrangement. Observe for significant cultural characteristics that might serve as an aid in identification. These include colony size and color (chromogenesis), Gram reaction, and hemolysis type.
2. Record your results in the table provided.

42 Dental Caries Susceptibility

Objectives

The student will be able to:

1. relate the acid production of certain mouth bacteria to dental caries susceptibility.
2. name the primary microorganism linked to dental caries.
3. list the significant ingredients of Snyder test agar.

Dental caries result when certain bacteria ferment carbohydrates on the tooth surface. The fermentation process produces lactic acid or other organic acids which decalcify the tooth enamel, allowing decay to begin. *Streptococcus mutans* has been considered to be the predominant organism responsible for dental caries. However, other organisms found as part of the normal mouth flora also produce acids and can contribute to the carious process.

The *Snyder test* is a simple method of determining caries susceptibility by measuring the rate of acid production from the metabolism of glucose by mouth microorganisms. As the microbes utilize the glucose present in the medium, the acids produced lower the pH below the uninoculated medium pH of 4.8. When the pH becomes 4.4 or lower, the pH indicator bromcresol green turns yellow. The susceptibility of an individual to caries is determined by the time it takes for the medium to turn yellow.

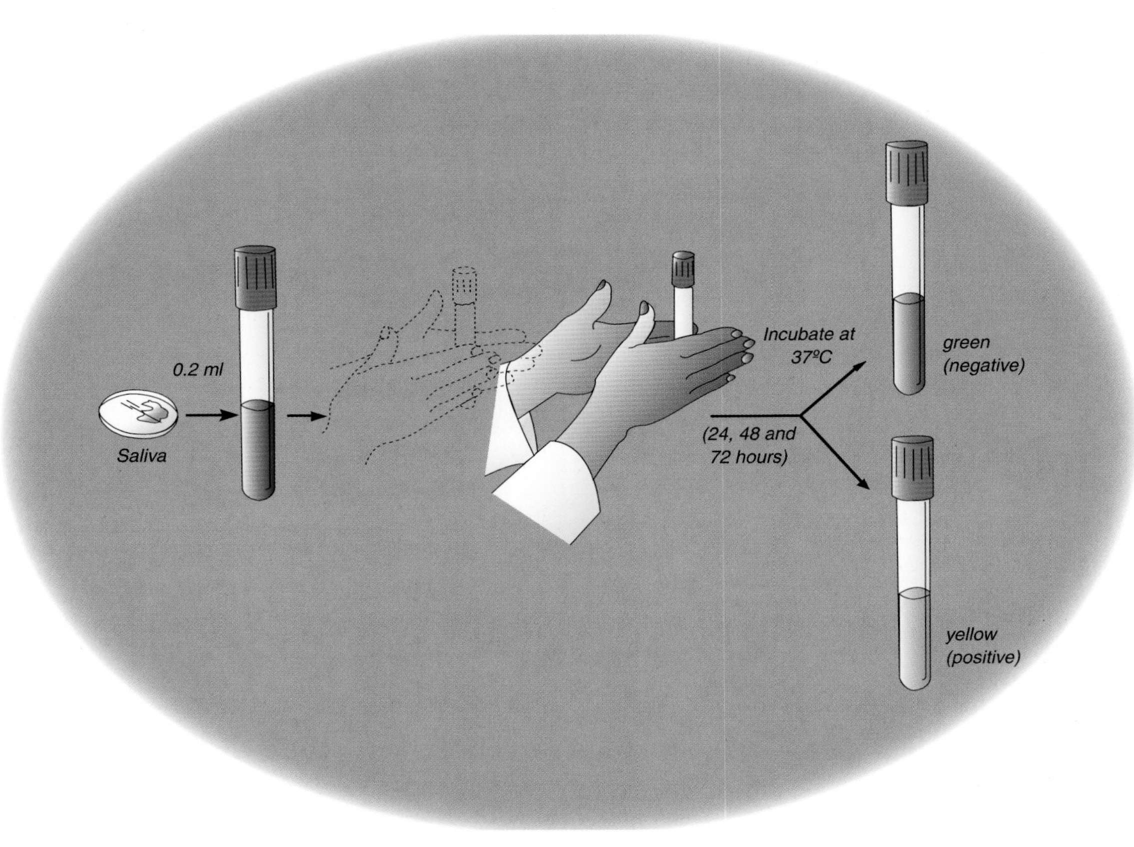

Figure 42-1

Snyder's dental caries susceptibility test

Table 42-1

Caries susceptibility scale

Caries susceptibility	Medium turns yellow in		
	24 hr	48 hr	72 hr
Marked	+		
Moderate	–	+	
Slight	–	–	+
None	–	–	–

Materials

1. Snyder test agar deep, sterile (one tube is a control; may be only a few per class)
2. Paraffin cube (size about 3 cm^3)
3. 1 sterile Petri plate
4. 1 sterile 1 ml pipet and aid

Procedure

Period 1

1. Melt a Snyder test agar deep and cool to 45°C.
2. While the deep cools, allow a paraffin cube to soften under your tongue and then chew it for 3 minutes. Do not swallow your saliva. Collect all of the saliva over the 3-minute period in a sterile empty Petri plate. Discard the paraffin.
3. Vigorously stir the saliva sample with the tip of a sterile 1 ml pipet for 30 seconds to more evenly distribute the organisms.
4. Aseptically pipet 0.2 ml of saliva into the cooled Snyder test agar.
5. Roll the tube between the palms of your hands until the saliva is evenly mixed with the medium. Allow it to solidify in an upright position.
6. Incubate the tube at 37°C.

Observations

Periods 2–4

1. Examine after 24, 48, and 72 hours to see if the pH indicator has turned yellow. Compare the tube to the control set aside for that purpose. Study the following table to determine the degree of your dental caries susceptibility (Table 42-1). Record your results in the table provided.

43 Handwashing and Skin Bacteria

Objectives

The student will be able to:

1. define the terms *transient* and *resident* microbes relative to the skin and hands.
2. explain the importance of handwashing or scrubbing to medicine, food preparation, and general health.
3. describe the general characteristics of the transient and resident bacteria.

The skin of humans and other animals serves as a habitat for many microbes living in pores and secretions and on skin tissue. Most of these **resident** microbes are Gram-positive bacteria and yeasts that are harmless under normal conditions and that even offer some protection from possibly pathogenic organisms. Protection is afforded in a number of ways, such as direct competition and production of acids and other inhibitory metabolites. The skin, especially the hands, acquires a large number of organisms through contact with many environmental sources of microbes. These microbes are not normal to the skin and usually don't grow but are not destroyed by the resident bacteria—at least not immediately. These microbes are termed **transient** microbes and are relatively easily removed from the skin—in contrast to the resident forms, which are very difficult to remove and cannot be removed completely. The nature of these organisms is highly varied, including Gram-positive and Gram-negative bacteria, yeasts, molds, viruses, and parasites, depending on the source.

The transient bacteria are most important on the hands, since hands are so intimately involved in many of the bodily portals of entry and exit of pathogens. These bacteria are easily transferred to food, cuts, surgical openings, and directly to the mouth or other body opening. The Austrian physician **Ignatz Semmelweis** observed in 1846-47 that puerperal fever in women just after giving birth was related to medical students not washing their hands before examining the women. Nurses usually washed, and during a period when the students were absent and only the nurses present, Semmelweis noted that the puerperal fever rate declined. He established a policy requiring the medical students to scrub their hands before examining the patients; as a result the puerperal fever rate dropped from 12% to slightly more than 1% in one year. This and later observations led to the surgical scrub used today in medical practice.

The hands are important in other ways as well. Examples include the transfer of foodborne illness organisms such as those causing typhoid fever, shigellosis, and salmonellosis from feces after a bowel movement, viruses from the nose and mouth and intestine, and from food-to-food (e.g., *Salmonella* organisms). Hands can also transfer organisms directly to the mouth and may play a very important role in the transmission of certain respiratory and intestinal diseases.

This exercise is intended to introduce the student to resident and transient organisms and the importance of handwashing as a method of hygiene.

Materials (per group of 4)

1. 2 sterile washbasins with 1000 ml of sterile tap water
2. 2 sterile surgical scrub brushes
3. 2 sterile 1 ml pipets with aid
4. 1 flask plate count agar (PCA), sterile
5. 1 flask mannitol salt agar (MSA), sterile
6. 18 sterile Petri plates
7. Hand lotion
8. 70% alcohol
9. L-shaped glass rod

Procedure

Period 1

A. Scrubbing

1. The group is divided into the scrubber, the helper, and two plate makers.
2. The person assigned to wash removes a sterile scrub brush from its wrapping while the helper removes the protective lid from the first bowl. The scrubber then wets both hands and the brush in the bowl.
3. The scrubber scrubs first one hand then the other for 30 seconds each. Scrubbing includes the palms and the backs of the hands as far as the wrist bone, including the fingers and under the nails. The helper reads instructions

and times and directs the scrubber as needed.

4. At the end of the scrub, the pair assigned to plate now removes the bowl and makes the first set of plates (see Part B below).
5. The scrubber now moves to a running water tap and, with the aid of the helper lathers both hands with soap, scrubbing them with the same brush as used in step 3 for one minute each. The hands are scrubbed in the same manner as previously described. Finally, the hands are thoroughly rinsed.
6. The helper now takes the used scrub brush, hands a sterile one to the scrubber and then uncovers the second bowl.
7. The scrubber wets both hands in the bowl and repeats the scrubbing of step 3 except that each hand is scrubbed 1 minute. At the end of scrubbing, the bowl is removed by the plating team for the second set of plates. The helper provides the scrubber with some hand lotion.

B. Plating

1. For bowl No. 1, lay out 9 plates according to Figure 43-1. Label 3 plates of MSA 1 ml, 0.5 ml, and 0.1 ml, respectively. Label 6 plates PCA, 2 with 1 ml, 2 with 0.5 ml, and 2 with 0.1 ml.
2. Take the L-shaped glass rod, rinse it with 70% alcohol and flame it. When cool, hold the long part of the rod between the palms of your hands. Lower the shorter end into bowl No. 1. Roll the rod rapidly between the palms so that the lower part revolves back and forth mixing the water of the bowl. If available, a magnetic stir bar can be sterilized in the bowl and the water mixed by the magnetic stir unit.
3. Pipet the appropriate volume from bowl No. 1 into each labeled plate. Pour melted cooled agar into the proper Petri plates and allow them to solidify.
4. For bowl No. 2, repeat the set up as done in step 1 of Plating above. Mix, pipet, and pour plates as described above and illustrated in Figure 43-1.
5. Incubate all plates inverted at 37°C for 24 hours.

Observations

Period 2

1. Count colonies on all plates using the following criteria:
   a. MSA. Mannitol salt agar is very selective for staphylococci, which are usually resident bacteria. *Staphylococcus aureus,* however, can be either resident or transient. Count all colonies growing on this medium.
   b. PCA. This is a general-purpose medium that will grow most skin organisms whether resident or transient. Count all colonies on this medium.

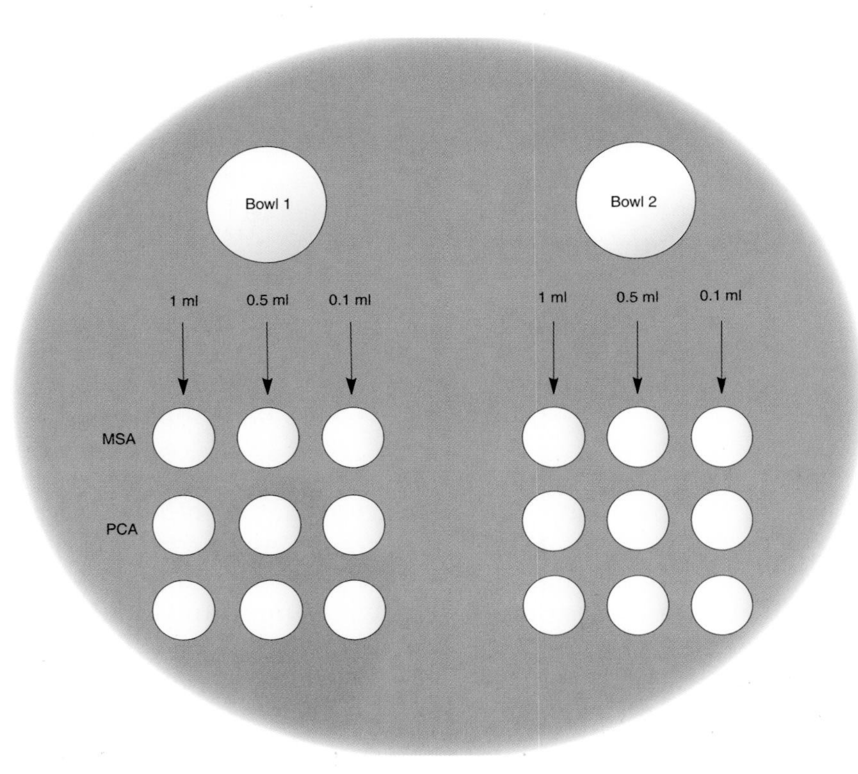

Figure 43-1

Plating design for handwashing procedure

Use the plate count criteria of Exercise 10 and make estimates according to the procedures presented there if plates are crowded.

2. Record counts in the table on the report form. Convert the count per ml of basin water to count per hand as follows:

$$\frac{\text{count/ml x 1000}}{2} = \text{count/hand}$$

3. Select 2 or 3 representative colonies on each plate and make Gram stains. Report the morphology and Gram reaction in the space provided on the report form.

Interpretation

Most people will have a considerable number of transient bacteria and some will have a few coliforms on the hands. These will be removed fairly easily. The PCA count should be significantly lower in the second bowl. The MSA counts in the second bowl may actually increase due to better removal. A third scrub would reduce the PCA count further but not as much as the first scrub. The third scrub should produce a lower MSA count than the second. MSA is highly selective for staphylococci and related bacteria.

44 Culture and Examination of Urine and Blood

Objectives

The student will be able to:

1. perform a culture of a urine specimen and/or a blood specimen.
2. name the differential and selective media used to cultivate organisms from urinary tract infections (UTIs).
3. explain the mechanism of the selective and differential capability of EMB agar.
4. name the bacterium that is the most common cause of UTIs.
5. name three organisms that are part of the normal urinary tract flora.
6. name the bacterium which is the most common cause of bacterial endocarditis.
7. perform a Gram stain of a clinical specimen and describe the bacteria present.
8. describe the decision-making progress performed by laboratory technologists when they evaluate and report on cultures from hospitalized patients.

Normal urine in the urinary bladder and the organs of the upper urinary tract are sterile. The **urethra**, however, does contain a normal resident flora that includes *Streptococcus* spp., *Bacteroides* spp., *Mycobacterium* spp., *Staphylococcus* spp., and *Neisseria* spp. Urine can become contaminated with these organisms as it passes through the urethra and is released.

The two main infections of the urinary tract are **cystitis** (inflammation of the urinary bladder) and **pyelonephritis** (inflammation of one or both kidneys). The most common cause of urinary tract infections (UTIs) is *Escherichia coli*. This organism usually causes an infection of the urinary bladder which may then spread up the ureters (ascending infection) to the kidneys to cause pyelonephritis. Other organisms capable of causing this type of infection are *Proteus* spp., *Pseudomonas* spp., *Enterobacter* spp., *Candida albicans*, and a few others.

Urinary tract infections are diagnosed by culturing a sample of urine on both blood agar, a good growth medium for many different organisms, and on **eosin-methylene blue (EMB) agar**, which is mildly selective for Gram-negative organisms.

Blood cultures are frequently obtained from patients who have developed a fever and are suspected of having bacteria in their bloodstream (blood is usually sterile). Approximately 5 ml of blood is collected at one drawing and added to 50 ml of trypticase soy broth (TSB), giving a 1:11 **ratio** of blood to broth. This dilution of blood is necessary to lessen the antibacterial factors in blood so the bacteria can grow easily.

While it was stated above that blood is normally sterile, there are periodic instances when transient bacteria appear briefly in the circulation. The most common bacteria are the **viridans group streptococci**. These alpha-hemolytic streptococci are the predominant flora of the mouth and enter the circulation through the gums during eating and chewing. Most people have no problem with eliminating these organisms. However, persons who have damaged heart valves (due to previous injury or congenital defect) are very susceptible to developing an infection of the damaged valves. The resulting disease develops slowly and is called **subacute bacterial endocarditis (SBE)**. This condition is diagnosed by the presence of intermittent fever, general weakness, a heart murmur, and repeated blood cultures that demonstrate growth of the infecting organisms shed from the infected valves.

In the hospital, culture plates of clinical specimens are examined after overnight incubation by either medical technologists or clinical microbiologists. It is at this initial viewing that the trained professional uses his or her **judgment** to determine the significance of the growth and what secondary steps should be taken.

The specimens you will culture yourself are illustrative of this decision-making process. A laboratory technologist would examine the urine culture and first decide if the growth represents normal flora colony types or if it is suggestive of the morphological types of bacteria known to cause urinary infections. If the bacterium does resemble a known urinary pathogen, is the amount of growth suggestive of an active infection or are there just a few colonies that could have resulted from fecal contamination of the specimen? If the amount and type of growth seen is suggestive of an infection, how can it be identified and does antibiotic susceptibility testing need to be performed? And last, the technologist must decide on how he/she should report the culture to the attending physician. This decision-making process requires good judgment and training.

The selective and differential media used in the clinical

microbiology laboratory are useful tools in identifying and processing microbial growth. **Eosin-methylene blue (EMB)** agar contains the dyes eosin-Y and methylene blue. These two dyes are inhibitory to the growth of many Gram-positive organisms and allow Gram-negative organisms to flourish without competition. The carbohydrate lactose is included in the medium to differentiate between organisms capable of lactose fermentation and those unable to ferment lactose. The latter types of organisms will form transparent colonies. *Escherichia coli*, which ferments lactose, not only forms colored colonies, but also produces a characteristic green metallic sheen (Figure 35-1). **Blood agar** is a very useful medium which allows uninhibited growth of most bacteria. This medium also helps to differentiate colony types based on their ability to cause hemolysis.

Materials (per pair)

1. 2 sterile 5% sheep blood agar plates
2. Eosin-methylene blue (EMB) agar plate, sterile
3. 2 cotton swabs, sterile
4. Clean microscope slides
5. Gram stain reagents
6. Urine specimen
7. Blood specimen

Procedure for Urine Specimen (2 days)

Period 1

1. Wet a sterile cotton swab in the urine specimen and inoculate a 1/4 section of a blood agar plate as illustrated in Figure 44-1. Repeat for an EMB agar plate. Return the swab to the urine specimen.
2. Streak both plates from the swabbed area for isolation using a sterile loop.
3. Incubate the plates at 37°C for 24 hours. These plates may be refrigerated after 24 hours if necessary to hold for a later laboratory period.
4. Using the swab in the urine specimen, transfer a large drop of urine from the container to a glass slide. Do not spread it around! Return the swab to the specimen for later disposal. Allow the drop to air dry (or use gentle heating) and then Gram stain.

Observations

Period 1

1. Observe the slide under oil immersion and observe cell morphology, cell arrangement, and Gram reaction. Record your results on the form provided. It may be necessary to search the

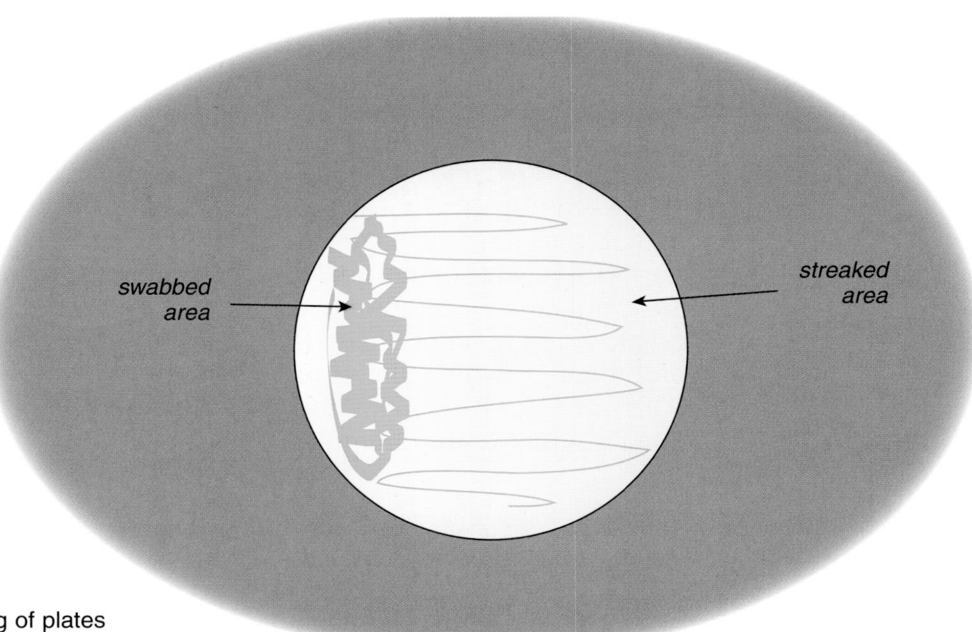

Figure 44-1

Swabbing and streaking of plates

slide extensively since there may be very few cells per field. Note that it takes about 500,000 cells per ml to have one cell per field with most oil immersion systems.

Period 2

2. Examine the urine culture plates and describe colony morphology, color, background reaction, if any, hemolysis, and any significant culture characteristics that might aid in identification.
3. Make Gram stains of representative colonies and note Gram reaction, morphology, cell arrangement, and any other useful characteristics. Record the results on the form provided.

Procedure for Blood Specimen (2 days)

Period 1

1. Wet a sterile cotton swab in the blood specimen and inoculate a 1/4 section of a blood agar plate as illustrated in Figure 44-1. Return the swab to the blood specimen.
2. Streak the plate from the swabbed area for isolation with a sterile loop.
3. Incubate the plates at 37°C for 24 hours. These plates may be refrigerated after 24 hours if necessary to hold for a later laboratory period.
4. Using the swab in the blood specimen, transfer a large drop of blood from the container to a glass slide. Do not spread it around! Return the swab to the specimen for later disposal. Allow the drop to air dry (or use very gentle heating) and then Gram stain.

Observations

Period 1

1. Observe the slide under oil immersion and observe cell morphology, cell arrangement, and Gram reaction. Record your results on the form provided. It may be necessary to search the slide extensively since there may be very few cells per field. Note that it takes about 500,000 cells per ml to have one cell per field with most oil immersion systems.

Period 2

2. Examine the urine culture plates and describe colony morphology, color, background reaction, if any, hemolysis, and any significant culture characteristics that might aid in identification.
3. Make Gram stains of representative colonies and note Gram reaction, morphology, cell arrangement, and any other useful characteristics. Record the results on the form provided.

Acknowledgment

We wish to thank Dr. Andy Anderson, Department of Biology, Utah State University, Logan, Utah, for permission to use this exercise.

45 Antimicrobial Susceptibility Testing

Objectives

The student will be able to:

1. perform an antimicrobial susceptibility disk assay.
2. measure a zone of inhibition.
3. distinguish between an antibiotic and a disinfectant.
4. describe the Kirby-Bauer test.
5. discuss the variation encountered in such tests as these.

Antibiotics by definition are chemical agents that are produced by living organisms, either killing or inhibiting the growth of other organisms. Antibiotics belong to a larger group of antimicrobial agents affecting growth and used in medicine called **chemotherapeutic agents**. Other chemotherapeutic agents, such as the sulfa drugs, are made chemically. Many of the antibiotics now can be made chemically or modified chemically so the distinction is blurred a bit. In this exercise, the term antibiotic will be used for all of these chemical agents. Antibiotics are generally distinguished from antiseptics and disinfectants (see Exercise 46) on a number of points. Antibiotics are generally effective in very small quantities and are usually very specific for one group of organisms. Among the antibiotics directed against prokaryotes, most are produced as secondary metabolites by three genera of microbes, *Bacillus* and *Streptomyces* from the bacteria and *Penicillium* from the fungi.

In clinical or hospital practice, the attending physician not only wishes to know the identity of an infecting organism but also wants to know the kind of chemotherapeutic agent to use in controlling the infection. The most effective agent, or better yet, the antibiotics that would be *ineffective*, can be determined by a simple laboratory test—the Kirby-Bauer test.

The **Kirby-Bauer method** of antibiotic sensitivity testing has evolved over a number of years to solve the problem of the many variations that are observed in such testing. The test is basically a diffusion test using a standard medium under standard conditions with a particular test organism. A paper disk is saturated with a known antibiotic concentration and placed in the center of a Petri plate previously heavily inoculated with a test organism. After incubation, any zone of inhibition (area of no growth) around the disk is measured and related to a standard for that compound and concentration. From this, the test organism can be said to be resistant, sensitive, or intermediate. No two tests are exactly alike, so standard concentrations are used. Variation occurs with such things as the volume of medium in the dish, the medium composition, the amount of inoculum, the solubility of the test agent, and many other factors. All of these factors must be controlled as much as possible. The standard medium for this method is **Mueller-Hinton agar**. These variations make comparison between different antibiotics difficult, but the test is useful for comparing different concentrations of the same antibiotic and, of course, differences between organisms. Standard strains of *Staphylococcus aureus* and *Escherichia coli* are used as controls for inhibitory zone diameter. The strains used in this exercise are not the standard strains but may be used in that manner.

This exercise is intended to introduce the student to a routinely used clinical laboratory procedure for determining antibiotic sensitivity.

Materials (per group of 4):

1. 18- to 24-hour-old broth cultures of *Staphylococcus aureus*, *Escherichia coli*, *Pseudomonas aeruginosa*, and *Saccharomyces cerevisiae*
2. Mueller-Hinton agar, 100 ml
3. 4 Petri plates
4. 4 sterile cotton swabs
5. Commercial antibiotic dispenser or individual disks
6. Forceps (if needed)
7. 70% alcohol (if forceps needed)
8. Calipers or plastic millimeter ruler

Procedure

Period 1

1. One student in the group of four pours four Petri plates of Mueller-Hinton agar. The agar should be the same depth in all the plates. Allow them to solidify.
2. Label each plate with the name of one of the four organisms assigned.
3. Each student then swabs one plate with the assigned organism. Aseptically remove a swab from its container and wet it in the

broth culture. Remove excess fluid by rolling the swab against the inside of the tube. Streak the swab uniformly across the entire surface of the plate, then rotate the plate 90° and repeat. Repeat the swabbing once more at a third angle. Discard the swab in an autoclave bag or the disinfectant solution provided. At the end of this step, each person in the group will have one plate inoculated with a different organism.

4. Disk placement: Use one of the following three methods:
   a. With a **commercial multiple disk dispenser**, remove the Petri plate top, place the dispenser over the agar surface and press firmly on the plunger. This results in depositing several disks onto the agar surface. Some devices press the disk down to ensure contact. If your device does not automatically tamp the disks, use flame-sterilized forceps to gently press each disk onto the agar surface. (Often these devices will not work if one dispenser is empty; check for empty holders beforehand.)
   b. With **individual dispensers**, release a disk one at a time around the margin of the Petri plate, about 1.5 cm inside the rim and about 2 cm apart.
   c. If **no dispenser** is available, transfer disks one at a time using sterilized forceps and gently press the disk onto the agar to ensure uniform contact.
5. Incubate the *S. cerevisiae* plate at 25°–30°C for 48 hours, and all others at 35°–37°C for 24 hours. Invert the plates.

Observations

Period 2 or 3

1. After incubation, observe each plate for zones of inhibition around the disks (i.e., a zone of no growth or clearing). Note whether partial growth inhibition occurs in the zone instead. Plates should be examined against a dark background, as illustrated in Figure 45-1.
2. Measure the diameter of the zone around each disk with calipers or a millimeter ruler to the nearest millimeter. Be sure the edge of the ruler bisects the center of the disk and the disk is included in the measurement. Record the name of the antibiotic and the measurement in the report form table. From Table 45-1, determine if your organism is resistant, intermediate, or sensitive to the antibiotic and enter an R, I, or S to the right of the measurement.
3. Record any unusual appearance: resistant colonies, concentric rings or bands of heavier growth, heavy growth ring at zone margin, precipitates in the zone or along its margin, etc.

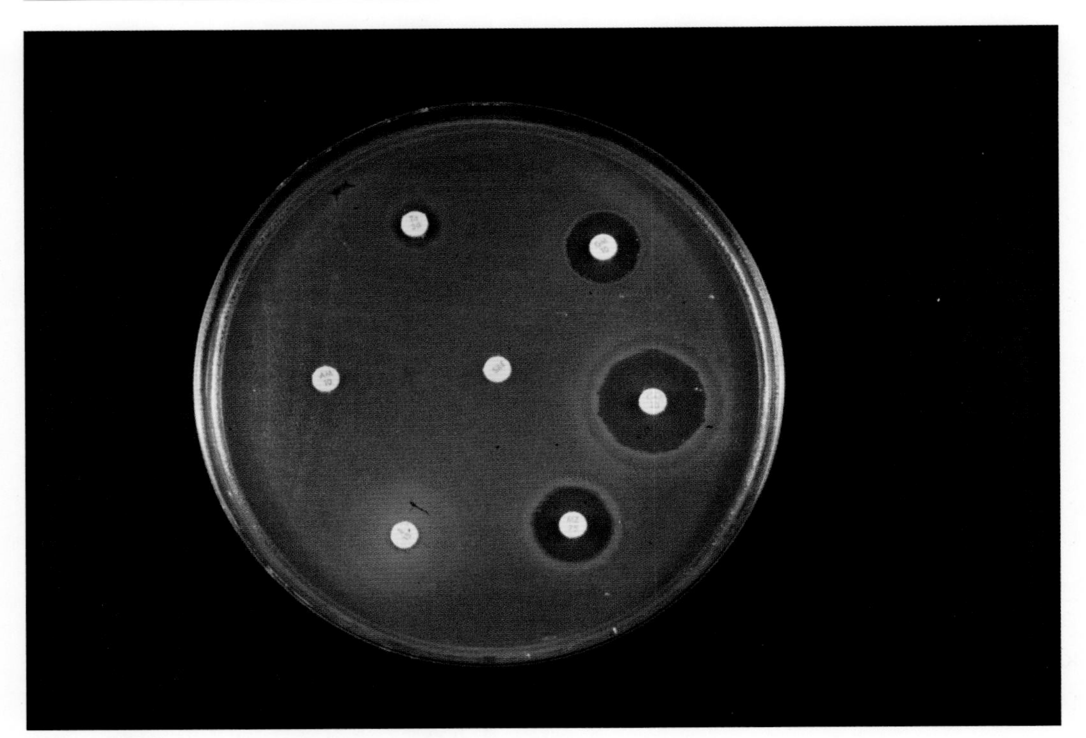

Figure 45-1

Example of antibiotic disks and zones of inhibition

Table 45-1

Antibiotic	Code	Conc.	Resistant	Intermediate	Sensitive
Ampicillin, Gram neg & enterococci	AM10	10 mcg	11	12–13	14
Ampicillin, staph & highly sensitive	AM10	10 mcg	20	21–28	29
Bacitracin	B10	10 U	8	9–12	13
Carbenicillin, Pseudomonas sp.	CB50	50 mcg	12	13–14	15
Cephaloglycin	CG30	30 mcg	16	17–26	27
Cephaloridine	CD30	30 mcg	11	12–15	16
Cephalothin	CF30	30 mcg	14	15–17	18
Chloramphenicol	C30	30 mcg	12	13–17	18
Clindamycin	CC2	2 mcg	11	12–15	16
Erythromycin	E15	15 mcg	13	14–17	18
Gentamicin	GM10	10 mcg	12		13
Kanamycin	K30	30 mcg	13	14–17	18
Lincomycin	L2	2 mcg	9	10–14	15
Methicillin	DP5	5 mcg	9	10–13	14
Nafcillin, Oxacillin	NF1, Ox1	1 mcg	10	11–12	13
Nalidixic acid	NA30	30 mcg	13	14–18	19
Neomycin	N30	30 mcg	12	13–16	17
Nitrofurantoin	F/M300	300 mcg	14	15–16	17
Novobiocin	NB30	30 mcg	17	18–21	22
Oleandomycin	OL15	15 mcg	11	12–21	17
Penicillin G, staph	P10	10 U	20	21–28	29
Penicillin G, other organisms	P10	10 U	11	12–21	22
Polymyxin B	PB300	300 U	8	9–11	12
Rifampin	RA5	5 mcg	24		
Streptomycin	S10	10 mcg	11	12–14	15
Triple sulfa	SSS.25	250 mcg	12	13–16	17
Tetracycline	Te30	30 mcg	14	15–18	19
Vancomycin	Va30	30 mcg	9	10–11	12

46 Action of Disinfectants and Antiseptics

Objectives

The student will be able to:

1. perform a disk plate inhibition assay for a disinfectant.
2. accurately measure a zone of inhibition.
3. distinguish between antiseptics, disinfectants, and sanitizers.
4. explain some of the reasons for variation in zone size between chemicals.

Antiseptics, disinfectants, and sanitizers are chemical agents used to kill microbes under different conditions. Antiseptics originally were agents that either killed pathogens or inhibited their growth, allowing the body's defenses to finish the job. These agents were used on skin or animate objects. Disinfectants had much the same connotation, except they were used only on inanimate objects. Today, **disinfectants** must kill vegetative cells but not endospores of bacteria and are commonly used on both animate and inanimate surfaces. Some viruses escape kill by disinfectants and the definition of disinfectant is currently under review. **Antiseptics** are defined essentially the same except for the inhibitory or bacteriostatic property. **Sanitizers** are chemical agents widely used in the food and restaurant industries. They are intended to kill a predetermined number of vegetative cells (not endospores) on an **already well cleaned** surface. Established public health standards dictate that less than 1 cell out of more than 1,000,000 can survive (i.e., a 99.9999+% reduction).

A number of tests for antiseptics and disinfectants are approved by the U.S. Food and Drug Administration, including the **Phenol Coefficient Test** (the effect of the agent on several standard pathogenic bacteria compared to phenol as a standard), the **Use Dilution Test** (the effectiveness of the agent under actual use conditions), and others. These are generally very demanding tests and require special equipment. In this exercise, a method is used that allows the student to see the inhibitory effects of the chemical. It is not a quantitative test, however, since each chemical agent behaves differently (see below). A standard base medium is poured into a Petri plate and the surface swabbed heavily with a test culture. A filter paper disk is saturated with the test chemical, drained, and placed on the agar surface in the center of the dish. After a period of incubation, any zone of inhibition is measured and recorded. A number of interesting additional observations can be made. Sometimes a ring of increased growth is seen at the rim of the inhibitory zone—a number of explanations have been proposed for this. Some agents cause mutations, and occasionally resistant colonies arise in the inhibitory zone with prolonged incubation.

This method, as with antibiotics, should not be used to compare these chemical agents directly. Each chemical is different in structure, solubility and thus diffusion, adherence to the filter paper disk, interaction with medium constituents, and the effect on the bacteria themselves. Similar kinds of chemicals behave in a similar manner, and dilutions of the same chemical are comparable.

Materials (groups of 4)

1. 18- to 24-hour-old broth cultures of *Staphylococcus aureus*, *Escherichia coli*, *Bacillus subtilis*, and *Saccharomyces cerevisiae*
2. 20 sterile Petri plates (5 for each organism)
3. 1 flask of nutrient agar (500 ml), sterile
4. 1 glass Petri plate with 25 paper disks, sterile
5. 4 sterile loosely wrapped cotton swabs
6. 70% alcohol
7. 5 chemical disinfectants, antiseptics, or sanitizers. Each student is encouraged to bring at least one (e.g., kitchen chemicals, mouthwashes, antiseptics, etc.)
8. 5 forceps (1 per chemical)
9. Calipers or plastic millimeter ruler

Procedure

Period 1

1. Each student is assigned to a group of four, one person to each available organism.
2. Each student pours 5 Petri plates from the flask provided and allows them to solidify. Label the cover with the name of the organism provided.
3. After the plates have solidified, aseptically remove a cotton swab from its container, and wet it in the broth culture assigned. Remove excess fluid by rolling the swab against the inside of the tube. Streak the swab uniformly across the entire

surface of the plate, then rotate the plate 90° and repeat. Repeat the swabbing once more at a third angle. Return the swab to the culture tube and repeat this procedure for **each** plate. Discard the swab in an autoclave bag or the disinfectant solution provided. At the end of this step, each student in the group will have five plates inoculated with one organism; four different organisms in the group.

4. Dip a forceps in alcohol and flame in the Bunsen burner. Allow the forceps to cool and pick up a filter disk. Dip the disk into the chemical to saturate it. Remove the disk from the chemical and touch it gently to the side of the container to draw off excess. Then place the disk in the center of one of your Petri plates. Label the Petri dish bottom with the **name** of the chemical and its **concentration**.
5. Repeat the previous step with each of your plates, each time using a **different** chemical. At the end of this procedure, each student will have one organism and 5 different chemicals.
6. Incubate the *S. cerevisiae* plates at 30°C for 24–48 hours. Incubate the remaining plates at 35°–37°C for 24 hours.

Observations

Period 2 or 3

1. After incubation, observe each plate for a zone of inhibition (i.e., a zone of no growth or clearing) around the disk. Note if only partial growth inhibition occurs or if well-separated colonies occur in the zone of inhibition. Plates should be examined against a dark background as in Exercise 45 (Figure 45-1).
2. Measure the diameter of the zone around the disk with the calipers or with a millimeter ruler to the nearest millimeter. Be sure the edge of the ruler bisects the center of the disk and the disk is included in the measurement. Record the measurement in the report form table. Record any unusual appearance: resistant colonies, concentric rings or bands of heavier growth, heavy growth ring at zone margin, precipitates in the zone or along its margin, etc.

47
Agglutination Reactions: Blood Grouping and the Rh Factor

Objectives

The student will be able to:

1. explain the agglutination principles involved in blood typing and the Rh factor determination.
2. describe the procedures performed to determine a person's blood type and Rh factor.
3. name the four most common agglutinogens associated with blood groups.
4. give the derivation of the symbol Rh.

Caution: *The United States Public Health Service has indicated caution is to be observed when working with human blood, since it can be a source of the AIDS virus. Since you are working only with your own blood, this exercise will be no hazard to yourself. Take great care to avoid contaminating the environment or others with your blood (or you with theirs). Discard all blood-contaminated materials exactly as instructed. Prepared reagents may be supplied instead. As an option, your instructor may provide an artificial or mock blood product as a substitute for human blood.*

The determination of blood group provides a good example of an antigen-antibody agglutination reaction in which large visible clumps are produced. In the ABO blood grouping system, there are two **antigens** (agglutinogens) that may be found in human red blood cells: A and B. A person having antigen A in his or her red cells is said to belong to group A; a person having antigen B belongs to group B; a person having both antigens belongs to group AB; and a person having neither antigen belongs to group O. In addition to these two antigens, blood cells have a large number of other antigens that can be tested for in much the same manner.

Whatever antigen a person has in his or her red blood cells, the corresponding **isoantibody** (agglutinin) is lacking in the serum. This obviously must be so since if a person of group A had the antibodies against A in the serum, the red cells would be **agglutinated** (clumped). When an antigen is not present, the corresponding isoantibody is present. Thus, a person of group A has no antibody A;

Blood type	Antigenic type of red blood cell (agglutinogens)	Antibodies in plasma or serum (agglutinins)
A	A	B
B	B	A
AB	AB	none
O	none	A and B

but not having antigen B, does have antibody B. A type AB person has neither antibody, while type O persons have both.

Blood typing is performed by adding commercially available antisera containing high **titers** of anti-A and anti-B agglutinins to suspensions of red blood cells. If agglutination (clumping of red blood cells) occurs only in the suspension to which the anti-A serum was added, the blood type is A. If agglutination occurs only in the anti-B mixture, the blood type is B. Agglutination in both samples indicates that the blood type is AB. The absence of agglutination indicates that the blood is type O.

In 1940 it was discovered that rabbit sera containing antibodies for the red blood cells of the Rhesus monkey would agglutinate the red blood cells of 85% of Caucasians. This antigen in humans, first designated as the *Rh factor*, was later found to exist as six antigens which were given the letters C, c, D, d, E, and e. Of these six antigens, the D factor is responsible for the Rh-positive condition. Determination of the Rh factor is accomplished by the addition of high titer commercial antisera containing anti-D agglutinins to a suspension of red blood cells. Agglutination indicates the Rh-positive condition.

Materials

1. Anti-A, anti-B, and anti-D human antisera
2. 2 clean microscope slides
3. 70% isopropyl alcohol
4. Cotton balls
5. Blood lancet
6. Applicator sticks or toothpicks
7. Disinfectant solution for used slides or pan for autoclaving them
8. If the instructor chooses to supply commercial red blood cells, items 3, 4, and 5 may be omitted.

Procedure and Observations

A. ABO Blood Typing

1. Divide a slide in half. Label one end A and the other end B.
2. Clean the tip of your finger with a cotton ball soaked in 70% isopropyl alcohol. Allow the alcohol to evaporate.
3. Aseptically unwrap the blood lancet and make a quick stab into the skin about 1-2 mm deep.
4. Wipe off the first drop of blood. Squeeze the finger and allow a drop or two of blood to fall on each end of the slide as in Figure 47-1.
5. Place one drop of anti-A typing serum on the end of the slide marked A.
6. Place one drop of anti-B typing serum on the end of the slide marked B.
7. Immediately mix each of the blood drops, first one and then the other, with a different applicator stick or a toothpick. Spread each mixture out to the size of a nickel and rock the slide and mixtures back and forth for 2 or 3 minutes.
8. Observe for the occurrence of agglutination (clumps) of red blood cells both macroscopically and microscopically (use the high-dry objective). Determine your blood type using the information included in the introduction to this exercise. Caution: Do not confuse drying up of the blood as agglutination.
9. Dispose of all lancets, cotton, applicators, or toothpicks *only* as instructed.

B. Rh Factor Determination

1. Place one drop of anti-D typing serum on a slide as in Figure 47-2.
2. Do a finger stick in the same manner as you did in the section on ABO blood typing.
3. Add two large drops of your blood to the antisera.
4. Mix thoroughly with an applicator stick or toothpick, spreading the mixture over most of the slide. Rock the slide gently back and forth for a period not to exceed 2 minutes.
5. Discard all blood-contaminated materials exactly as instructed.
6. Observe for macroscopic agglutination. Determine your Rh factor with agglutination being positive and no agglutination being negative.

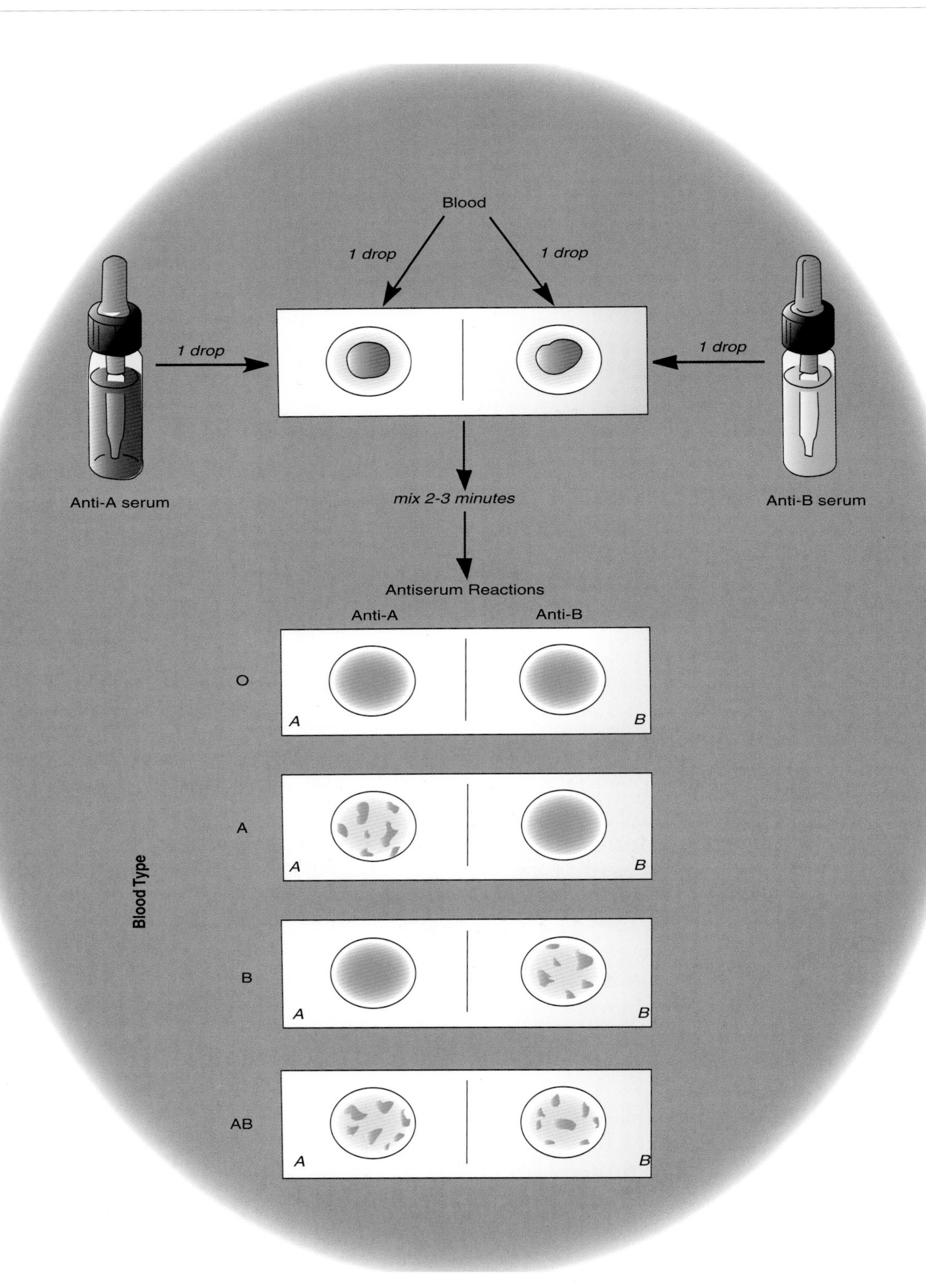

Figure 47-1

ABO blood typing reactions

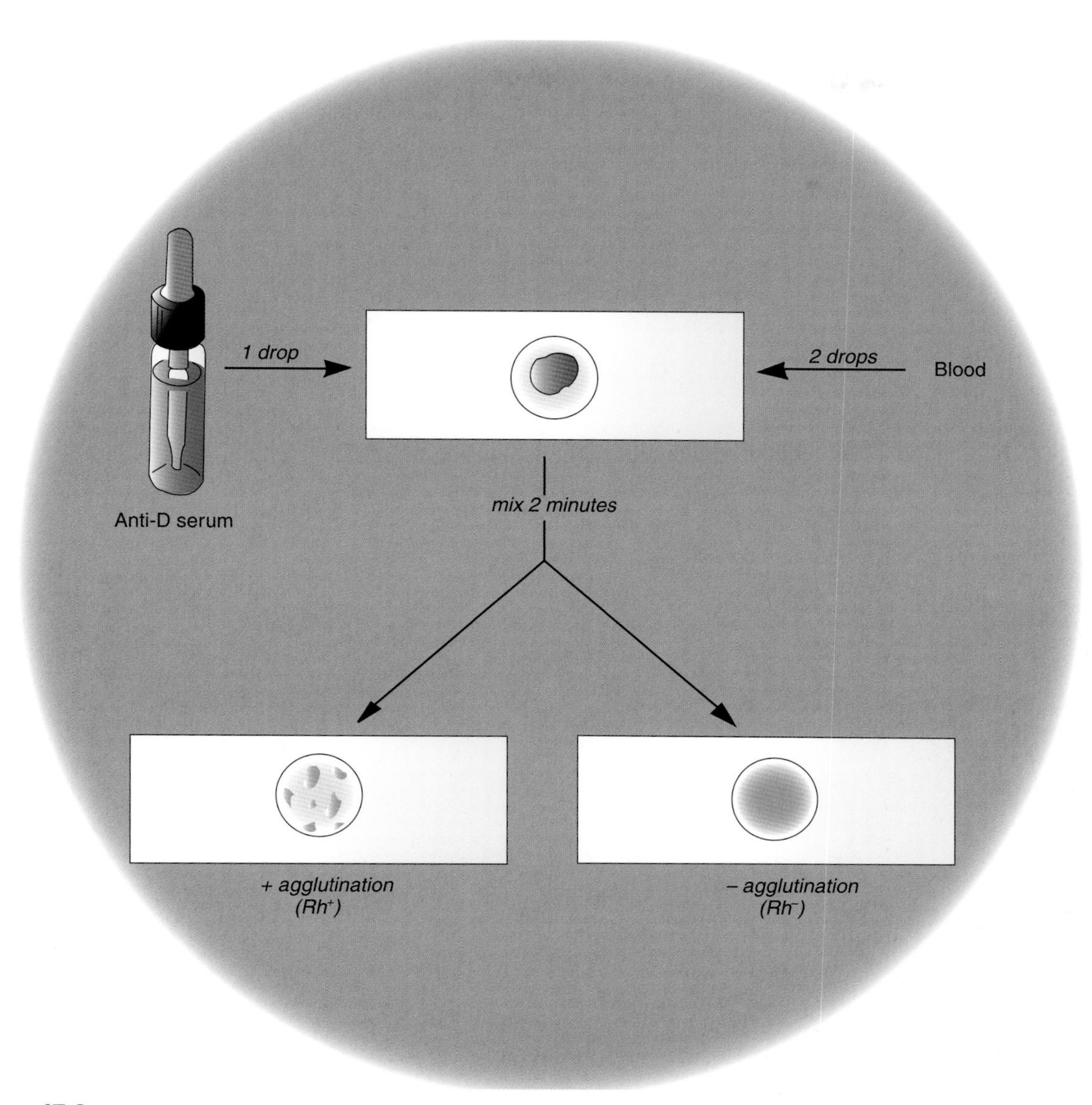

Figure 47-2

Rh factor determination

48 Enzyme-Linked Immunosorbent Assay (ELISA)

Objectives

1. perform an indirect enzyme-linked immunosorbent assay.
2. describe the theory of enzyme immunoassays in general.
3. diagram the procedure of the antibody assay used in this exercise.

Agglutination and precipitation reactions are excellent rapid procedures for serological determinations, but they are limited in sensitivity, usually in the 1–100 ng range. By coupling the antibodies or antigens to fluorochromes, radioisotopes, or enzymes, the detection sensitivity can be increased to the picogram range. Enzymes covalently conjugated chemically to antibody or antigen allow the detection of antigen-antibody reactions indirectly through the formation of enzyme product, often colored for easy detection and spectrophotometric quantitation. These are referred to as **Enzyme Immunoassays** or **EIA**. When the antigen or antibody is adsorbed to a surface, such as beads or microtiter wells, it is referred to as an **Enzyme-Linked ImmunoSorbent Assay** or simply **ELISA**. Several ELISA methods are available. In this exercise you will do the indirect method (Figure 48-1) in which an **antigen** is adsorbed onto the walls of microtiter wells, washed, and followed with a specific **antibody** prepared in rabbit. These are allowed to react, the excess antibody washed away, and a **rabbit globulin antibody urease** conjugate is added. These react, and the excess is washed away. **Urea substrate** containing bromcresol purple at pH 4.8 (yellow color) is added. If retained on the well walls by antigen, the enzyme conjugated to the antibody hydrolyzes the urea substrate, releasing ammonia which causes the bromcresol purple to change to its alkaline color, a deep purple. Thus color indicates the presence of enzyme, in turn indicating the presence of the antibody to rabbit globulin and the antigen-antibody complex. The reciprocal of the highest dilution at which the purple color first appears is the antibody titer. Several other enzymes (e.g., horseradish peroxidase, alkaline phosphatase, and beta-galactosidase) can also be used.

Materials (per pair)

1. 5 ml chicken egg albumin solution (10 μg/ml) in adsorption buffer
2. 1.0 ml of chicken egg albumin antiserum (from rabbit) diluted 1:50 in dilution buffer in a 13 x 100 mm tube
3. 5 ml antiserum to rabbit globulin (from goat) conjugated with urease diluted 1:500 in dilution buffer
4. Wash buffer in a plastic squeeze bottle
5. 5 ml urea-bromcresol purple substrate
6. Microtiter plate, flat bottom
7. 2 sterile 1 ml pipets and safety aid
8. 5 sterile 0.2 ml pipets
9. Parafilm to cover the microtiter plate
10. Beakers or other containers to receive washings
11. ELISA plate reading spectrophotometer, if available (not necessary)
12. Distilled water

Procedure

Period 1

1. Add 0.2 ml of egg albumin antigen (10 μg/ml) to microtiter plate wells A1 to A10 and A12, skipping A11.
2. Add 0.2 ml of adsorption buffer to wells B1 to B10 and B12, skipping B11. This row serves as a control. Cover the plate with parafilm and label.
3. Incubate the plate for 1 hour at room temperature and then at 4°C overnight or until the next lab period to allow the albumin antigen to adsorb to the well walls.

Period 2

4. Label the tube of the egg albumin antiserum (from rabbit) No. 1. Make 2-fold dilutions of the egg albumin antiserum (Figure 48-2) as follows:
   a. Add 0.5 ml of dilution buffer to *each* of nine 13 x 100 mm tubes. Label them 2 through 10.
   b. Using the same pipet, transfer 0.5 ml of serum from tube 1 to tube 2 and mix with the pipet.
   c. With the same pipet, transfer 0.5 ml from tube 2 to tube 3 and mix with the pipet.
   d. Continue in like manner through tube 10 and discard the pipet.
5. Remove the parafilm (save it) from the plate and invert the plate over a container to empty the wells.

6. Wash the wells by gently filling them with wash buffer from the plastic squeeze bottle. Cover with the parafilm and let the tray set for 3 minutes. Do this each time a wash is done.
7. Empty the wells into the container (save the parafilm). Repeat step 6 two more times.
8. Add 0.2 ml of the highest dilution of the serum to each of wells A10 and B10. With the same pipet, add 0.2 ml of the next lowest dilution to each of wells A9 and B9. With the same pipet, continue in this manner until the lowest dilution is added to wells A1 and B1.
9. Cover the plate with the parafilm and incubate at room temperature for 1 hour.
10. Wash all wells with the wash buffer as in step 6 above, allowing the plate to set covered 3 minutes. Wash at least three times in this manner.
11. Add 0.2 ml rabbit globulin antiserum (from goat) conjugated to urease (1:500 in dilution buffer) to every well.
12. Cover with the parafilm and allow to set at room temperature 1 hour.
13. Wash all wells at least 2 times as in step 6 with wash buffer followed by two similar washes with distilled water to remove all buffer that would interfere with the detection of ammonia.
14. Add 0.2 ml of urea-bromcresol purple substrate to each well. *Be careful   to touch the sides of the wells with the pipet.*
15. Cover with parafilm and incubate at room temperature for 1 hour.

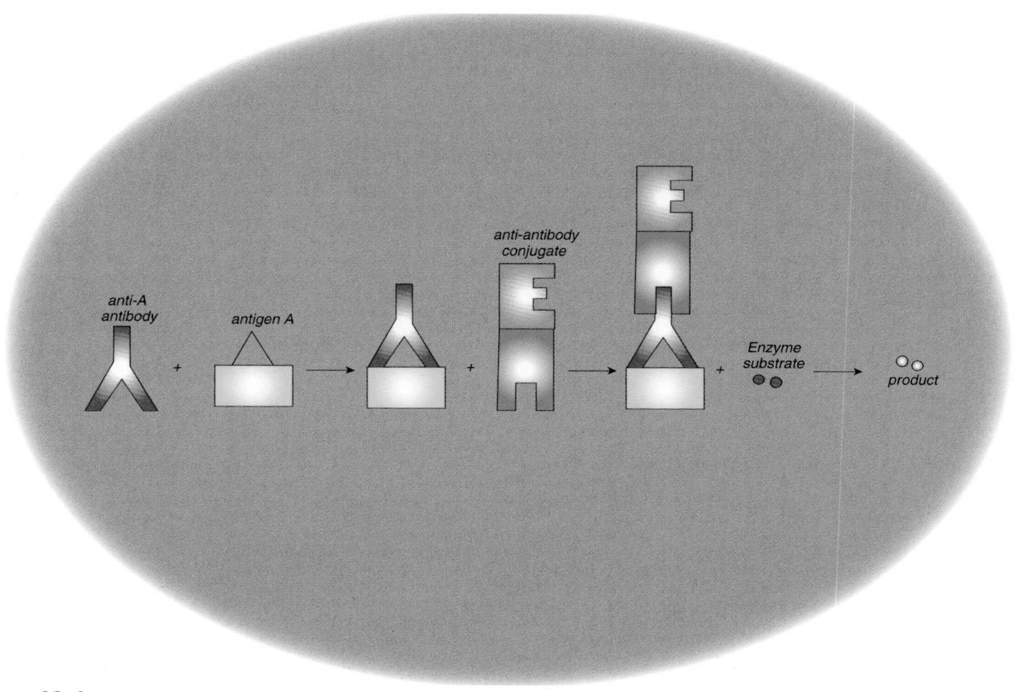

Figure 48-1

Enzyme-linked immunosorbent assay procedure

Observations

Period 2

1. Read the wells visually. A purple color indicates the presence of egg albumin antibody in the well. The reciprocal of the highest dilution giving a distinct purple color is the antiserum titer for this exercise. Enter the observations on the report form.
2. If an ELISA well reader is available, read at 588 nm using well A12 to zero the instrument and subtracting the absorbance of the corresponding row B (controls) well from each reading. An adjusted absorbance >0.1 is considered positive. The titer is calculated as in step 1.

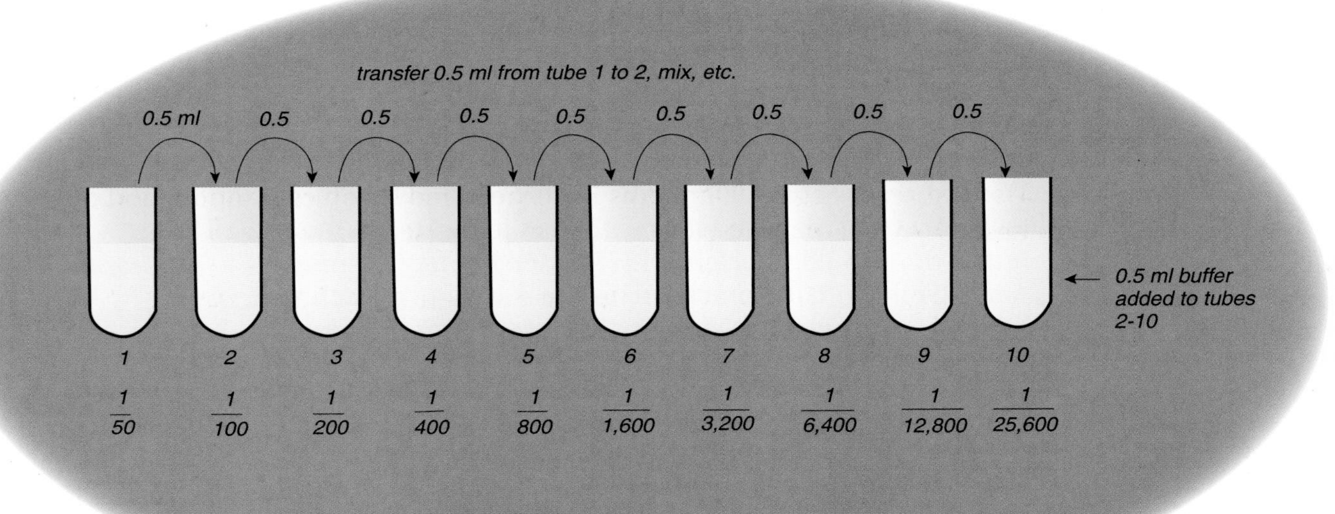

Figure 48-2

Antiserum dilution sequence

Appendix I
Media

All media except 1, 6, 8, 11, 13, 14, 20, 25, 27, 30, 31, 34, and 36 are available commercially. Numbers in parentheses following the media refer to the exercise in which it is used.

1. **Acetate agar**: glucose, 1.0 g; yeast extract, 2.0 g; Na acetate•$3H_2O$, 5.0 g; agar, 15 g; distilled water 1 L, pH 5.5. (3)

2. **Blood agar**: Use tryptic soy agar 100 ml plus 5 ml sheep red blood cells. Add blood only when agar has cooled to about 45°C. Plates can be purchased ready for use. (20,33,41,44)

3. **Brilliant green lactose bile (BGLB) broth**: Peptone 10 g; lactose 10 g; oxgall 20 g; brilliant green 0.133g; distilled water 1 L, pH 7.2. (35)

Casein agar: see skim milk agar

Dilution water: see peptone water 0.1%

4. **EC broth**: Tryptose 20 g; lactose 5 g; bile salts No. 3 1.5 g; K_2HPO_4 4 g; KH_2PO4 1.5 g; NaCl 5 g; distilled water 1 L, pH 6.9. (35)

5. **EMB agar** (eosin methylene blue agar): Peptone 10 g; lactose 10 g; K_2HPO_4 2 g; eosin Y 0.4 g; methylene blue 0.065 g; agar 15 g; distilled water 1 L, pH 7.1. (35,44)

6. **Gelatin nutrient agar**: Nutrient agar plus 10 g gelatin; 1 L distilled water. (17,33)

7. **Kligler iron agar**: Beef extract 3 g; yeast extract 3 g; peptone 15 g; proteose peptone 5 g; lactose 10 g; glucose 1 g; $FeSO_4$ 0.2 g; NaCl 5 g; Na thiosulfate 0.3 g; phenol red 0.024 g; agar 15 g; distilled water 1 L, pH 7.4 (22,33)

8. **LAB agar**: Lactose 20 g; tryptone 10 g; meat extract 10 g; yeast extract 10 g; tomato juice (filtered) 50 ml; Tween 80 1 g; K_2PO_4 2 g; agar 15 g; distilled water 1 L, pH 6.6. (38)

9. **Lauryl sulfate tryptose lactose (LST) broth**: Tryptose 20 g; lactose 5 g; K_2HPO_4 2.75 g; KH_2PO_4 2.75 g; sodium lauryl sulfate 0.1 g; distilled water 1 L, pH 6.8. (35)

10. **Litmus milk**: Skim milk 100 g; litmus 5 g; distilled water 1 L, pH 6.8. (18,33)

11. **Manganese agar**: Nutrient agar plus 6 mg/L $MnCl_2$. (14,33)

12. **Mannitol salt agar**: Proteose peptone No. 3 10 g; beef extract 1 g; D-mannitol 10 g; NaCl 75 g; phenol red 0.025 g; agar 15 g; distilled water 1 L, pH 7.4. (43)

13. **Marine broth**: Peptone 5 g; yeast extract 1 g; ferric citrate 0.1 g; NaCl 19.45 g; $MgCl_2$ (dried) 5.9 g; $NaSO_4$ 3.24 g; $CaCl_2$ 1.8 g; KCl 0.55 g; Na_2HCO_3 0.16 g; $SrCl_2$ 0.034 g; H_3BO_3 0.022 g; Na_2SiO_3 0.004 g; NaF 0.0024 g; NH_4NO_3 .0016 g; Na_2HPO_4 0.008 g; distilled water 1 L, pH 7.6. (For agar add 15 g per L.) (28)

14. **Minimal salts agar (M56)**: (39)
    Solution A: $Na_2HPO_4 \cdot 7H_2O$ 8.2 g; KH_2PO_4 2.7 g; $(NH_4)_2SO_4$ 1 g; $FeSO_4 \cdot 7H_2O$ 0.25 mg; distilled water 1 L, pH 7.2. Cool to 50°C.
    Solution B: 10% $MgSO_4 \cdot 7H_2O$
    Solution C: 0.5% $Ca(NO_3)_2$
    Solution D: 20% glucose
    Solution E: 4% L-methionine (8% DL-form)
    Solution F: 4% L-leucine (8% DL-form)
    Solution G: 4% L-threonine (8% DL-form)
    Solution H: 4% L-alanine (8% DL-form)
    Solution I: 3.2% agar (32 g/L)
    Autoclave all solutions separately. To 1 L of solution A, aseptically add 1 ml of B, 1 ml of C, and 10 ml of D. Then add 1 ml of the required amino acids, temper to 50°C, and mix 1:1 with solution I, also tempered to 50°C. Pour in labeled plates. The final concentration of each amino acid in the agar plate is 20 mg/ml.

15. **Motility test medium**: Tryptose 10 g; NaCl 5 g; agar 5 g; distilled water 1 L, pH 7.2. (32)

16. **MR-VP medium**: Buffered peptone 7 g; glucose 5 g; K_2HPO_4 5 g; distilled water 1 L, pH 6.9. (23,33)

17. **Mueller-Hinton agar**: Beef infusion 300 g; casamino acids (technical) 17.5 g; starch 1.5 g; agar 17 g; distilled water 1 L, pH 7.3. (45,46)

18. **Nitrate broth**: Beef extract 3 g; peptone 5 g; KNO_3 1 g; distilled water 1 L, pH 7.0. (25,33)

19. **Nutrient broth and agar**: Beef extract 3 g; peptone 5 g; distilled water 1 L, pH 6.8. Add 15 g/L agar for the agar medium. (6,7,8,9,10,14,15,16,17,18,19,21,22,23,24,25,26,27,29, 31,32,33,34,35,45,46)

20. **Peptone broth or agar**: Peptone 8 g; beef extract 1 g; agar 15 g; distilled water 1 L. (39)

21. **Peptone water** (0.1% or 4%): Peptone 1 g per 1 L or 4 g per L distilled water. (10,26,33,36,37,38)

22. **Phenol red broth**: Trypticase or proteose peptone No. 3 10 g; beef extract (optional) 1 g; phenol red (7.2 ml of 0.25% solution) 0.00018 g; distilled water 1 L, pH 7.3. Carbohydrates 5 g per L broth added before autoclaving. Autoclave 10 min. at 118°C. (21,33)

23. **Plate count agar** (Standard plate count agar): Yeast extract 2.5 g; tryptone 5 g; glucose 1 g; agar 15 g; distilled water 1 L, pH 7.0. (33,36,37,43)

24. **Sabouraud's dextrose or maltose agar/broth**: Polypeptone or neopeptone 10 g; dextrose (or maltose) 40 g; agar 15 g; distilled water 1 L, pH 5.6 before autoclaving. For broth: before melting, filter the agar out using Whatman filter paper. (3,45)

25. **Salt agar**: Table salt (NaCl) 200 g (for 20% w/v); $MgSO_4 \cdot H_2O$ 20 g; KCl 5 g; $CaCl_2 \cdot 6H_2O$ 0.2 g; yeast extract 3 g; tryptone 5 g; distilled water 1 L, pH 7.2–7.4. (29)

Sheep blood agar: see blood agar

26. **Simmons citrate agar**: Sodium citrate 2 g; NaCl 5 g; K_2HPO_4 1 g; $NH_4H_2PO_4$ 1 g; $MgSO_4 \cdot H_2O$ 0.2 g; bromthymol blue 0.08 g; agar 15 g; distilled water 1 L, pH 6.8. (23,33)

27. **Skim milk nutrient agar** (20%): Skim milk powder 20 g; 100 ml distilled water. Autoclave. Mix 20 ml sterile skim milk with 100 ml previously sterilized, melted, nutrient agar, cooled and held at 50°C. (17,33)

28. **Snyder test agar**: Tryptose 20 g; glucose 20 g; NaCl 5 g; bromcresol green 0.02 g; agar 20 g; distilled water 1 L, pH 4.8. (42)

29. **Spirit blue agar**: Tryptone 10 g; yeast extract 5 g; spirit blue 0.15 g; agar 20 g; distilled water 1 L, pH 6.8. Sterilize. Use lipase reagent (Difco) or prepare lipid as follows: warm 400 ml distilled water, add 1 ml of Tween 80, add 100 ml of olive oil (or other lipid) and homogenize vigorously. Add 30 ml of the homogenate to 1 L sterile, melted and cooled to 50°C agar. (16,33)

30. **Starch agar**: Nutrient agar plus 10 g per L of soluble starch. (15,33)

31. **Toluidine blue-DNA agar (TB-DNA)**: DNA 0.3 g; $CaCl_2$ (anhydrous) 1.1 mg; NaCl 10 g; Tris (hydroxymethyl) aminomethane 6.1 g; agar 10 g; toluidine blue O 0.083 g; distilled water 1 L. (19)
Dissolve Tris in 1 L distilled water and adjust to pH 9.0. Add remaining ingredients except toluidine blue O and heat to boiling. After agar and DNA are dissolved, add the toluidine blue O and dispense in smaller quantities. **DO NOT** autoclave. Medium is stable at room temperature for about 4 months and is satisfactory even after several remelting cycles.

32. **Tryptic soy agar**: Tryptone 17 g; soytone 3 g; NaCl 5 g; agar 15 g; distilled water 1 L, pH 7.3. (7,11,12,13,24,33)

33. **Tryptic soy broth**: Tryptone 17 g; soytone 3 g; NaCl 5 g; glucose 2.5 g; K_2HPO_4 2.5 g; distilled water 1 L, pH 7.3. (7,18,20,28,30,33)

34. **Tryptone broth** (1%): Tryptone 10 g; distilled water 1 L, pH 7.0. (23,33)

35. **Urea broth**: Urea 20 g; yeast extract 0.1 g; KH_2PO_4 9.1 g; Na_2HPO_4 9.5 g; phenol red 0.0001 g (or 4 ml of a 0.25% aqueous solution); distilled water 1 L, pH 6.8. **FILTER STERILIZE. DO NOT AUTOCLAVE.** (27,33)

36. **Yeast extract tryptone agar**: Yeast extract 2.5 g; tryptone 5 g; agar 7.5 g; distilled water 1 L, pH 7.0. (33)

Reference: Difco Manual, 10th edition. Difco Laboratories, Detroit, MI 48232. 1984.

Appendix II
Reagents

Note: Some one-use reagents are given in the Instructor's Manual for the specific exercise (e.g., Exercise 48).

1. **Acid-alcohol**: 95% ethyl alcohol plus 3% (v/v) concentrated HCl. Dissolve the HCl in the alcohol.

2. **Acidified mercuric chloride**: $HgCl_2$ 15 g; distilled water 100 ml; concentrated HCl 20 ml. Mix in this order. (Note: Trichloroacetic acid or 4 N HCl may be used instead.)

Barritt's reagents: see Voges-Proskauer A and B

3. **Carbolfuchsin** (Ziehl-Neelsen): Solution A - basic fuchsin 0.3 g; 95% ethyl alcohol 10 ml. Solution B - phenol 5 g; distilled water 95 ml. Mix solutions A and B and let stand for several days before use. Filter through paper into stock bottle.

4. **Ferric chloride reagent**: $FeCl_3$ 10 g; distilled water 100 ml.

5. **Gram stain reagents**:

   Gram's crystal violet: Solution A - crystal (gentian) violet 2 g; 95% ethyl alcohol 20 ml. Solution B - ammonium oxalate 0.8 g; distilled water 80 ml. Mix solutions A and B and store 24 hours. Filter through paper into stock bottle.

   Gram's iodine: Dissolve KI 2 g in 300 ml distilled water. When completely dissolved, add I_2 1 g.

   Acetone-alcohol: 95% ethyl alcohol 80 ml; acetone 20 ml.

   Safranin: Safranin O (2.5% in 95% ethyl alcohol) 10 ml; distilled water 100 ml.

6. **Hydrogen peroxide (3%)**: 30% H_2O_2 10 ml; add distilled water to make 100 ml. Other concentrations calculated accordingly.

7. **Indole reagent** (Kovac's): Para-dimethyl-aminobenzaldehyde 5 g; amyl or butyl alcohol 75 ml; concentrated HCl 25 ml. Dissolve the reagent in the alcohol, warming gently in a 37°C water bath. When completely dissolved, add the HCl carefully while stirring.

Kovac's reagents: see indole and oxidase reagents

8. **Loeffler's methylene blue**: Solution A - methylene blue 0.3 g; 95% ethyl alcohol 30 ml. Solution B - KOH 0.01 g; distilled water 100 ml. Mix solutions A and B and filter through paper.

9. **Malachite green** (5%): Malachite green 5 g; distilled water to 100 ml. Filter through paper.

10. **Methyl cellulose**: Carboxymethylcellulose (methocel), 15 centipoise (Sigma Chemical Co., St. Louis, MO). Mix 10 g in 45 ml boiling water, cool, and add 45 ml water.

11. **Methyl red**: Methyl red 0.2 g; 95% ethyl alcohol 500 ml; distilled water 500 ml. Filter if necessary.

12. **Nitrite A**: Sulfanilic acid 0.8 g; acetic acid 5N (1 part glacial to 2.5 parts water) 100 ml.

13. **Nitrite B**: N,N' dimethyl-1-naphthylamine 0.5 g; acetic acid 5N (see nitrite A) 100 ml. (The base may be dissolved in alcohol instead.)

14. **Nessler's reagent**: Dissolve 50 g KI in 35 ml cold distilled water. Add a solution of saturated mercuric chloride until a slight precipitate persists. Add 400 ml of a 50% solution of KOH. Add sufficient distilled water to make 1 L, allow to settle, and decant the supernatant for use.

15. **Oxidase reagent** (Kovac's): Tetramethyl-para-phenylenediamine HCl 0.5 g; distilled water 50 ml. Prepare the solution on the day of use. If this is not possible, refrigerate to store but warm to room temperature before use. Do not use if more than 5 days old or if a precipitate has formed.

16. **Tris-maleic acid buffer**: 2-amino-2-(hydroxymethyl)-1, 3 propanediol (Tris) 6 g; maleic acid 5.8 g; $(NH_4)_2SO_4$ 1 g; $FeSO_4 \cdot 7H_2O$ 0.25 mg; distilled water 900 ml, pH 6.0. Autoclave at 121°C for 15 minutes. Cool to 50°C and aseptically add 100 ml of sterile $MgSO_4 \cdot 7H_2O$ 1 g and $Ca(NO_3)_2$ 50 mg in 100 ml distilled water.

17. **Vaspar**: Equal parts Vaseline and paraffin. Sterilize in the oven at 160°C for 3 hours. Avoid browning. Melt in boiling water bath. Open flames can ignite the vapors.

18. **Voges-Proskauer A** (Barritt's A): Alpha naphthol 5 g; 95% ethyl alcohol.

19. **Voges-Proskauer B** (Barritt's B): KOH 40 g; creatine 0.3 g; distilled water 100 ml.

Ziehl-Neelsen carbolfuchsin: see carbolfuchsin

Appendix III
Microorganism List

American Type Culture Collection Preceptrol cultures are suitable in most cases. Exceptions are indicated below. A number of supply houses sell suitable cultures for most exercises.

An asterisk (*) identifies species used only in Exercise 33 as unknowns and may be omitted if not used. Parentheses refer to the exercise in which microorganisms are used.

Fungi

1. *Aspergillus* sp. (3)
2. *Penicillium* sp. (3)
3. *Rhizopus* sp. (3)
4. *Saccharomyces cerevisiae* (3,29,30,33,45,46)

Bacteria

5. *Acinetobacter calcoaceticus* (18,33)
6. *Alcaligenes faecalis* (21,33)
7. *Aquaspirillum serpens** (33)
8. *Arthrobacter globiformis** (33)
9. *Bacillus* sp. ATCC 21592 (30)
10. *Bacillus cereus* (12,14,17,33)
11. *Bacillus polymyxa** (33)
12. *Bacillus stearothermophilus* (28)
13. *Bacillus subtilis* (7,15,21,25,26,32,33,46)
14. *Chromobacterium violaceum** (33)
15. *Enterobacter aerogenes* (20,23,33)
16. *Enterococcus faecalis* (7,18,20,24,30,33)
17. *Escherichia coli* (7,8,9,10,11,12,15,16,17,18,21,22,23,24,27,28,29,30,33,44,45,46)
18. *Escherichia coli* K12 ATCC 25404P (39)
19. *Flavobacterium aquatile** (33)
20. *Halobacterium salinarium* (29)

21. *Lactococcus lactis** (33)
22. *Micrococcus luteus* (7,8,9,33)
23. *Micrococcus roseus** (33)
24. *Mycobacterium smegmatis* (13,33)
25. *Proteus vulgaris* (22,27,33)
26. *Pseudomonas aeruginosa* (16,18,24,25,26,32,33,45)
27. *Pseudomonas fluorescens I* (28,33)
28. *Pseudomonas fluorescens IV** (33)
29. *Salmonella enteriditis (typhimirium)** (33)
30. *Serratia marcescens* (7,17,19,20,31,33)
31. *Staphylococcus aureus* (12,13,14,19,21,24,25,29,30,31,32,33,45,46)
32. *Staphylococcus epidermidis* (19)
33. *Streptococcus mitis* ATCC 9811 (44)
34. *Vibrio marinus* ATCC 15381 (28)

Name ______________________________ Date _______ Grade _________

1. Care and Use of the Microscope

Questions

A. Fill in the Blanks

1. If a lens is in sharp focus on a specimen and an adjacent lens is swung into position with need for only minor adjustment in the focus, the lens is said to be ______________________.

2–3. When a lens is rotated to a higher power, the diameter of the field your eye sees when looking through the microscope is (larger/smaller/the same), but the actual size of the specimen as measured on the slide is (larger/smaller/the same).

4. Total magnification of a lens system is calculated by ______________ the magnification of each lens in the system.

5. The microscope should be stored with the ______ ______ lens in the down position.

6. The resolving power of a lens depends directly on the ____________________ of light used.

7. The distance between the specimen and the objective when it is in focus is called the __________ ____________ p.

8–9. Numerical aperture is directly related to the __________ of __________ of the glass in a lens.

10. A lens with pH inscribed on the side would be found on a __________ __________ microscope.

B. Multiple Choice

Select the best answer for the following statements.

_____ 11. If the total magnification with a 45X high-power objective is 225X, what would be the magnification of the ocular?
a. 5X
b. 10X
c. 12X
d. 15X

_____ 12. The resolution of a microscope is increased by:
a. using a shorter wavelength of light
b. decreasing the amount of light emerging from the diaphragm
c. using a condenser with low numerical aperture
d. removing the condenser from the microscope

C. Short Answer

13. Given: Blue light of wavelength 400 nm, a condenser, and an objective with a numerical aperture of 1.2. What is the smallest resolvable object in mm diameter seen with such a microscope? Show work.

14. Describe how immersion oil helps in viewing an object. A diagram in the space below may be helpful in answering this question.

D. Matching

Match the parts of the microscope with their functions.

____ 15. Ocular (eyepiece)
____ 16. Revolving nosepiece
____ 17. Objective
____ 18. Diaphragm
____ 19. Condenser
____ 20. Mechanical stage
____ 21. Base
____ 22. Arm
____ 23. Coarse adjustment
____ 24. Fine adjustment
____ 25. Body tube

a. Opens and closes with a lever controlling the amount of light striking the object
b. A series of lenses that usually magnify 10 times
c. Condenses light waves into a cone, thereby preventing escape of light rays; raised and lowered to control amount of light striking object
d. Raised and lowered in focusing some microscopes
e. The lens closest to the object
f. Supports upper portion of microscope
g. Rotates to change from one objective to another
h. Moves stage or body tube up and down rapidly for purposes of approximate focusing
i. Allows the slide to be moved
j. Supports entire microscope
k. Moves stage or body tube up and down very slowly for purposes of definitive focusing

E. Completion

26–34. Complete the following table with regard to your microscope.

Objective	*Objective Magnification*	*Ocular Magnification*	*Total Magnification*	*N.A.*	*Resolution*
Low-power					
High-power					
Oil immersion					

Name ______________________ Date ________ Grade ________

2. Bacteria and Cyanobacteria

Results and Observations

Bacterium
Name __________
Morphology __________
Color __________

Bacterium
Name __________
Morphology __________
Color __________

Bacterium
Name __________
Morphology __________
Color __________

Bacterium
Name __________
Morphology __________
Color __________

Bacterium
Name __________
Morphology __________
Color __________

Yeast
Name __________
Morphology __________

Cyanobacterium
Name __________
Morphology __________

Cyanobacterium
Name __________
Morphology __________

Cyanobacterium
Name __________
Morphology __________

If more observations are necessary, use the above format on a separate sheet of paper.

Questions

1. Name one bacterium from this exercise that is stained red or pink.
   ____

2. Name one bacterium from this exercise that is stained purple. ____________________

3. Give the morphological name for each of the following bacterial shapes:

   a. spheres irregularly clustered ________________________

   b. spheres in a chain ________________________

   c. spiral S shape ________________________

   d. one-half a spiral turn ____________________

   e. cells dividing remaining side by side ________________

   f. a cylinder shape ________________________

4. Name a reproductive structure found in some cyanobacteria. _______________

5–8. Name the four genera of cyanobacteria suggested for use in this exercise.

   5. ____________________________ 6. ____________________________

   7. ____________________________ 8. ____________________________

9. The function of a heterocyst is to ____________ ______________ .

10. Heterocysts are found in ______________________ (name group).

11. If you observe a microbe that is 2 μm long and 1 μm in diameter, what is its volume?
    __________μm^3

12. If a photograph was taken through the microscope of a cell measuring 1.3 μm long and the photograph image was 13 mm long, what is the magnification of the photographic image? ________X

Name ______________________________ Date ________ Grade ________

3. Fungi

Results and Observations

I and/or II. Prepared slides and/or fungal slide culture drawings

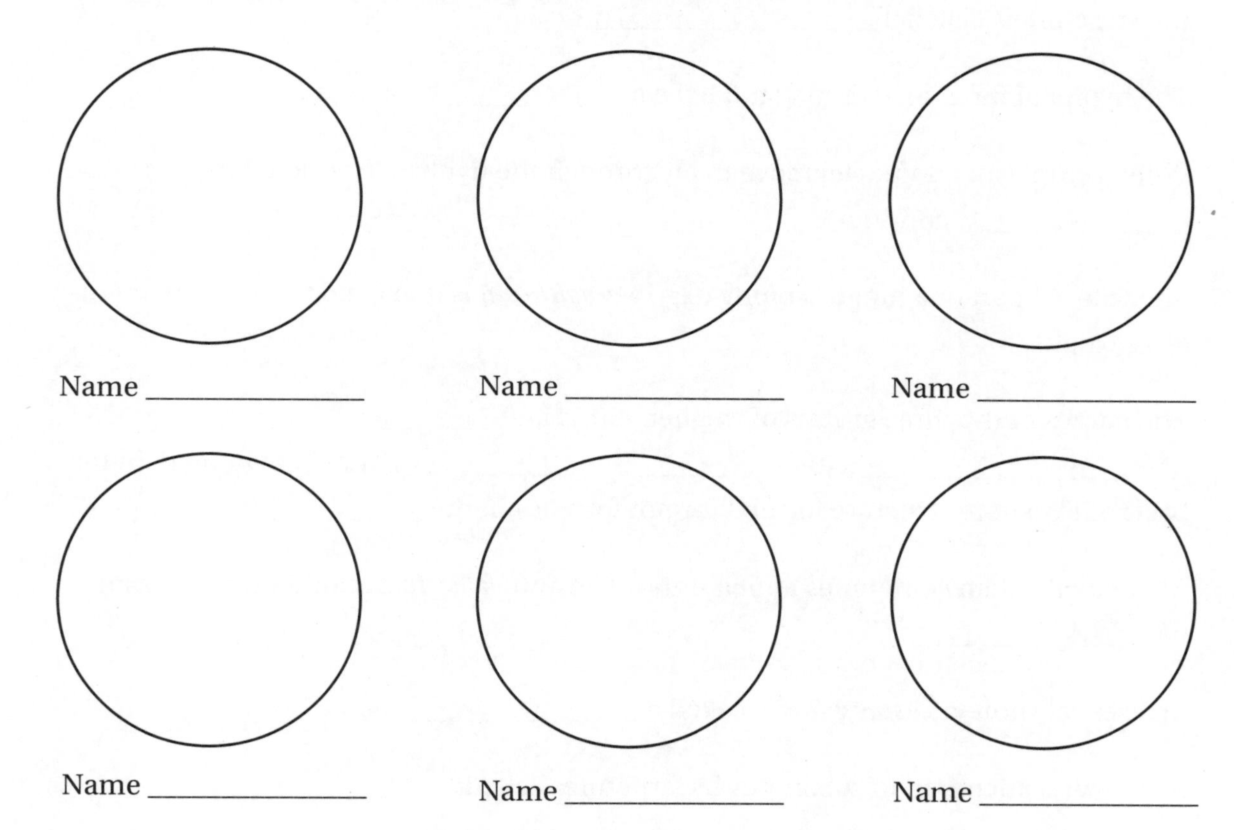

III. Yeast wet mount or prepared slide drawings

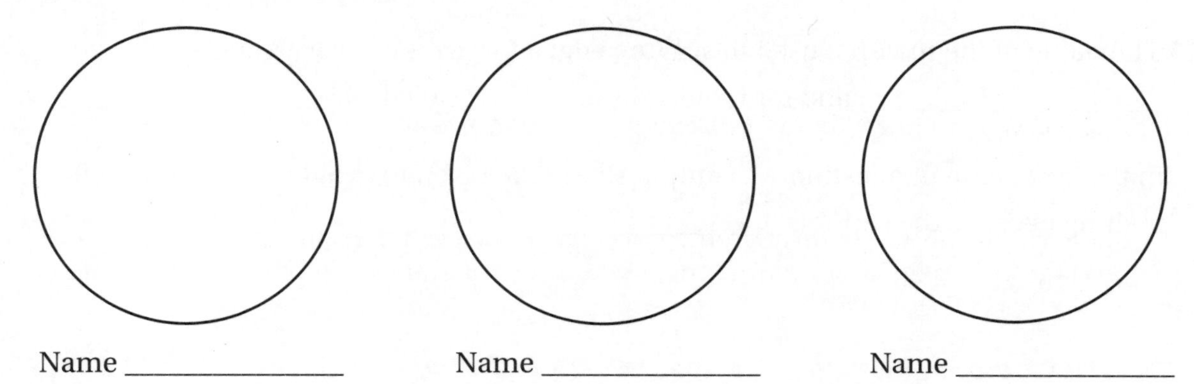

Questions

1–3. The tubular part of fungus growth is called a ________________ . After much branching, the mass of tubes is called a ________________ . When the mass of tubes is visible to the eye as a cottony growth the tubes are commonly called a ________________ .

4. If a fungus growing underground produces an above-ground sexual fruiting structure, the structure is called a ________________ .

5. The hypha of fungi may or may not have a ______________ across it.

6. When cytoplasm and nuclei move freely through the hyphae, it is called the ________________ condition.

7. The colored part of a fungus colony (e.g., *Penicillium*) is due to spores that are (sexual/asexual).

8–11. The names of the three groups of "higher" fungi are ________________ , ________________ , and ________________ . These are named on the basis of the spore structure formed (or not formed) in the ______________ cycle.

12. A fungus that can sometimes appear mold-like and yeast-like at other times is said to be ________________ .

13. An asexual spore of *Aspergillus* is called a ________________ .

14. *Rhizopus* is anchored to substrates by structures called ________________ .

15. In the yeast *Saccharomyces cerevisiae*, both haploid and diploid cells reproduce by ________________ .

16–17. The name of the material used to seal the edge of wet-mount coverslips is ________________ , and for mold culture slides is called ________________ .

18. In the latter case in question 16, caution should be taken to avoid overheating while melting because the vapor is ______________ .

Name ____ Date ________ Grade __________

4. Protozoa and Algae

Results and Observations

1. With the aid of the diagrams below, place a check mark next to each structural element as you observe it. You may not be able to see all of the structures in a particular specimen.

2. Other microscopic observations

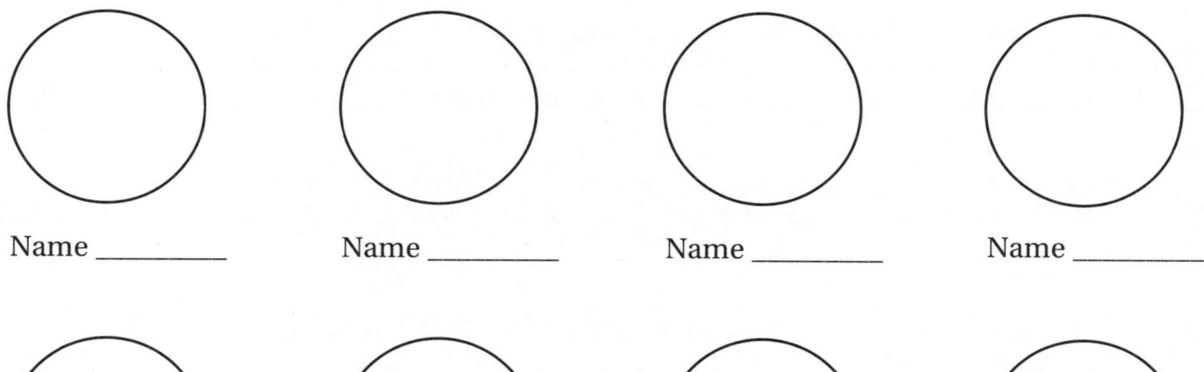

Name ________ Name ________ Name ________ Name ________

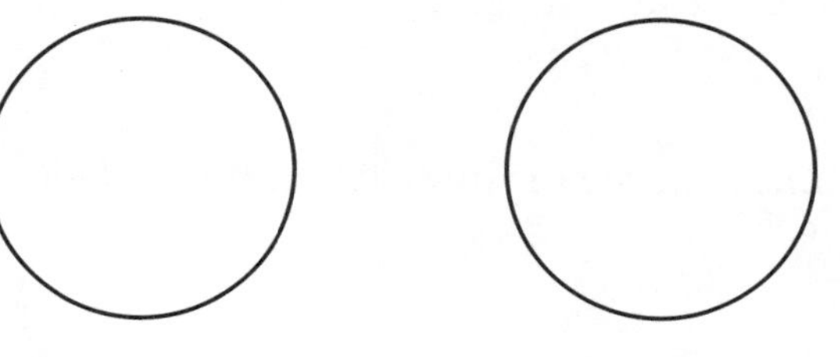

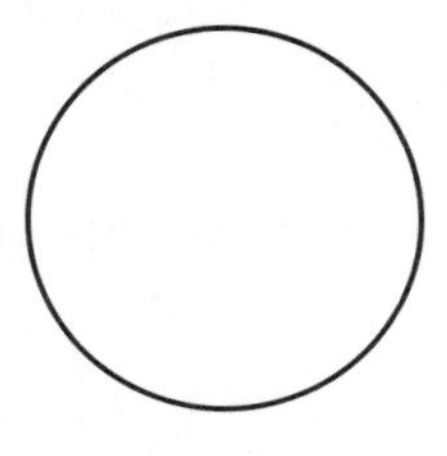

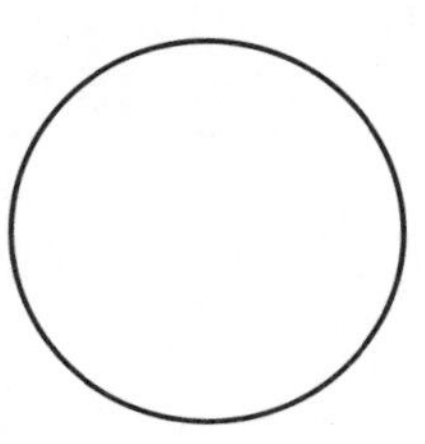

Name ________ Name ________ Name ________ Name ________

Questions

1. The common name of those organisms with a silica cell wall is ________________ .

2. A simple chemical test for starch in cells is the use of ________________________ .

3. The red algae have the accessory pigments called ______________ .

4. The most commonly found secondary chlorophyll among the algae would appear to be ________________ .

5. The nucleus of the protozoa and algae and the chloroplast of the algae are surrounded by a unit __________________ .

6. You would expect to find the flagellated algae classified among the ________________________ as protozoa.

7–10. Name a genus of the diatoms _________________ , the euglenids _______________ , the dinoflagellates ________________ , and the green algae __________________ .

11–13. Name a parasite in the group Mastigophora _________________ , Sarcodina ________________ , and Sporozoa ___________________ .

14. The chemical used to slow motility of algae and protozoa is ____________________ .

15. In which group of algae is gliding motility found? ______________ .

16. Which group of protozoa is entirely parasitic? _______________ .

17. To what group does *Amoeba proteus* belong? __________________ .

18–19. The malaria life cycle requires two hosts, man and a _________________ , usually of the genus __________________ .

20. *Giardia lamblia* is an intestinal parasite causing diarrhea. It is transmitted to humans in __________ .

21. *Trypanosoma gambiense* causes the human disease known as _____________ ____________ ____________ .

22. A very common cause of vaginitis in the human female or urethritis in the male is __________________ .

23. A filamentous green alga is __________________ .

Name ______________________________ Date ________ Grade ________

5. Helminths

Results and Observations

Schistosoma mansoni

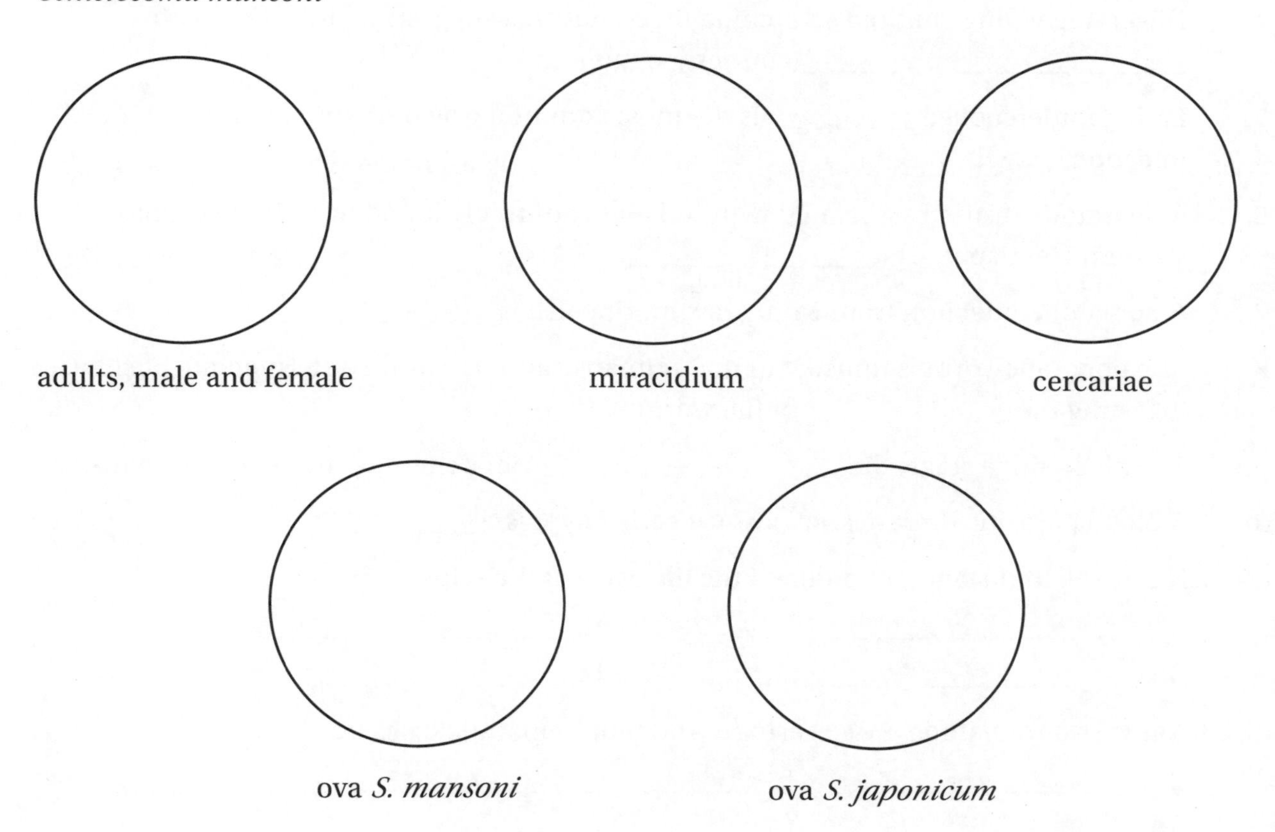

adults, male and female　　miracidium　　cercariae

ova *S. mansoni*　　ova *S. japonicum*

Disease ______________________________

Using the above format, prepare drawings for the following on separate sheets of paper:

Clonorchis sinensis: adult, ova
Taenia solium: scolex, ova, cysticerous
Taenia saginata: proglottid
Diphyllobothrium latum: scolex, ova
Enterobius vermicularis: adult, ova
Necator americanus: adult, ova
Trichinella spiralis: encysted larvae, migratory larvae

Questions

1. The intermediate host from which humans acquire *Clonorchis sinensis* is raw ________________ ______________ .
2. A scolex is characteristically found in ____________________ .
3. The fish tapeworm is acquired by eating raw ________________ (common fish name).
4. Observing young children scratching their anal area suggests an infection with __________________________ (common name).
5. Eating undercooked __________ is the most common way humans acquire trichinella infection.
6. A nematode disease associated with soil and commonly found in Africa and the Western Hemisphere is ____________________ .
7. A nematode infection from eating raw marine fish is ____________________ .
8. The pork tapeworm is unusual in that humans can serve as the intermediate host by ingesting the ______________ of the worm.
9. The blood fluke stage, the ____________________, can penetrate the skin of humans.
10. A blood parasitic disease known since early Egypt is ____________________________ .

11–14. Name the four stages, in order, of the life cycle of the schistosomes:

11.__________________________ 12.______________________

13.__________________________ 14.______________________

15–23. Name two trematode, three cestode, and four nematode diseases:

15.____________________ 16.____________________ 17.____________________

18.____________________ 19.____________________ 20.____________________

21.____________________ 22.____________________ 23.____________________

24–30. Define the following terms:

24. monoecious
25. dioecious
26. proglottid
27. flatworm
28. cestode
29. cysticercus
30. scolex

Name ______________________________ Date ________ Grade ________

6. Media Preparation and Sterilization

Results and Observations

1. Describe the appearance of the three broth tubes:
   a. Sterile tube

   b. Room temperature tube

   c. Refrigerated tube

   What reason can you give for the difference between the refrigerated and room temperature tubes?

   What would you predict would happen if you continued to store the refrigerated tube in the refrigerator for several weeks or more?

2. Describe the appearance of the agar beaker medium left at room temperature.

   Where did these organisms come from?

   To what do you attribute the change in appearance of the agar slant after the agar solidifies?

Questions

1. A medium consisting of glucose, ammonium chloride, and a few mineral salts would be best called a __________________ medium.

2. Adding a pH indicator to a medium would make it a ________________ medium.

3–4. A medium containing human blood would be called a __________________ medium unless it was also used to separate the reactions of two organisms, in which case it also would be a ______________________ medium.

5. A nutrient required by a bacterium in extremely small amount such as molybdenum would be considered a _________-nutrient.

6. Adding an inhibiting dye to nutrient agar makes a ________________ medium out of it.

7. Agar comes from a type of _________________ .

8. Agar contains a polysaccharide which, in addition to carbon, nitrogen, and hydrogen, also contains the element _____________ .

9–10. Agar melts at _______°C and solidifies at about _______°C.

11. Room temperature is about ________°C.

12. Standard agar media contain about ___% agar.

13. Boiling of a medium (will/will not) sterilize it.

14. When an object is rendered free of living organisms, it is called ___________________ .

15–16. Autoclaving is routinely done at ______°C under ________ of pressure.

17. A household utensil that can accomplish the same sterilizing effect as an autoclave is a ____________ ____________ .

18. The process of handling sterile media after autoclaving is referred to as ______________ _______________ .

19. In order to sterilize media in volumes larger than 1 liter, it may be necessary to increase the autoclaving _______________ .

20. The boiling point of water at an elevation of 1500 meters (about 5000 feet) is (lower/higher) than at sea level.

21. The medium called motility test medium contains ____% agar.

Name ______________________________ Date __________ Grade __________

7. Broth and Agar Slant Culture

Results and Observations

Part A: Broth Culture[a]

Bacterial Species	Control	*Serratia marcescens*	*Bacillus subtilis*	*Enterococcus faecalis*	*Micrococcus luteus*	*Escherichia coli*
Sketch of growth						
Type of surface growth*						
Amount of turbidity*						
Amount of sediment*						
Chromogenesis/ consistency						

[b]Refer to Figure 7-3
*None = 0; Slight= +; Moderate = ++; Heavy = +++

Bacterial Species	Control	*Serratia marcescens*	*Bacillus subtilis*	*Enterococcus faecalis*	*Micrococcus luteus*	*Escherichia coli*
Sketch of growth						
Amount of growth*						
Form of growth*						
Chromogenesis/ consistency						

[b]Refer to Figure 7-4
*None = 0; Slight= +; Moderate = ++; Heavy = +++

Questions

Part A

1. The small amount of growth on the end of an inoculating loop is called an ___________________ .
2. A flame-sterilized inoculating ________ is used to inoculate the medium.
3. Aseptic technique is important to prevent self-infection and __________________________ of the culture medium.
4. In sterilizing your transfer loop, the ________________ end must also be sterilized.
5. Setting the culture cap down on the bench is a violation of ____________ _____________ .
6. When growth on or in a medium contains only a single species of microorganism, it is called a __________ ______________ .
7. A term often used to describe cloudiness in broth media is ____________________ .
8. When a bacterial culture grows across the surface of a broth, the growth is referred to as a ______________ .
9. Cells that settle to the bottom of a broth tube are referred to as ________________ .
10. Growth occurring as clumps suspended in the broth is referred to as ________________ .

Part B

11. Growth around the margin of the surface of a broth tube but not across it is referred to as a _______________________.
12. Name one of the ingredients in nutrient broth: ______________________ .
13. An isolated population of bacteria growing on a solid medium is called a ____________ .
14. A slant culture of an organism which is used to make other subcultures is called a ____________ culture.

15–17. List three factors that might influence the cultural characteristics of an organism growing on an agar medium:

15. ______________________________________
16. ______________________________________
17. ______________________________________

18. A word meaning "pigmentation" or "color" used to describe growth on an agar slant or colony is __________________________ .
19. The term used when a culture is placed at a particular temperature for a period of time is _____________________ .
20. When an inoculated tube is compared to an uninoculated tube for signs of difference, the uninoculated tube serves as a ____________ .

Name ______________________________ Date ________ Grade ________

8. Streak Plate

Results and Observations

Observation	Bacterial species	
	Micrococcus luteus	*Escherichia coli*
Sketch of colonies		
Size (mm dia.)		
Form		
Elevation		
Margin		
Chromogenesis[a]		
Light passage[b]		

[a]Pigmentation

[b]Translucent: light passes through but cannot read print through colony

Transparent: can read print through colony

Opaque: no light passes through colony

Questions

1. If asepsis is not practiced in the laboratory, ____________________ of cultures will result.
2. As suggested in the introduction, bacterial growth means an increase in bacterial ____________________ instead of the size of one cell.
3. A good aid in determining the margin type of a bacterial colony is a ________________ lens.
4. When you incubate a Petri plate, it should normally be placed in an incubator with the ____________ side up.
5. A mass of cells growing on an agar surface is called a ______________ .
6. The mass of cells in question 5 is usually derived from a ___________ cell.
7. Restreaking a colony for isolation several times is necessary to be reasonably sure that you have a _________ ____________ .
8. In taking cells from a culture tube for streaking, it is important to ______________ the end of the inoculating loop holder before entering the tube.
9. Streaking a plate with the open face of the loop oriented vertically to the agar surface is a good way to ______________ the agar.
10. Agar should be poured into a plate at a fairly cool temperature to avoid excess ____________________ on the inside of the lid.

Name ______________________________ Date ________ Grade ________

9. Pour Plate

Results and Observations

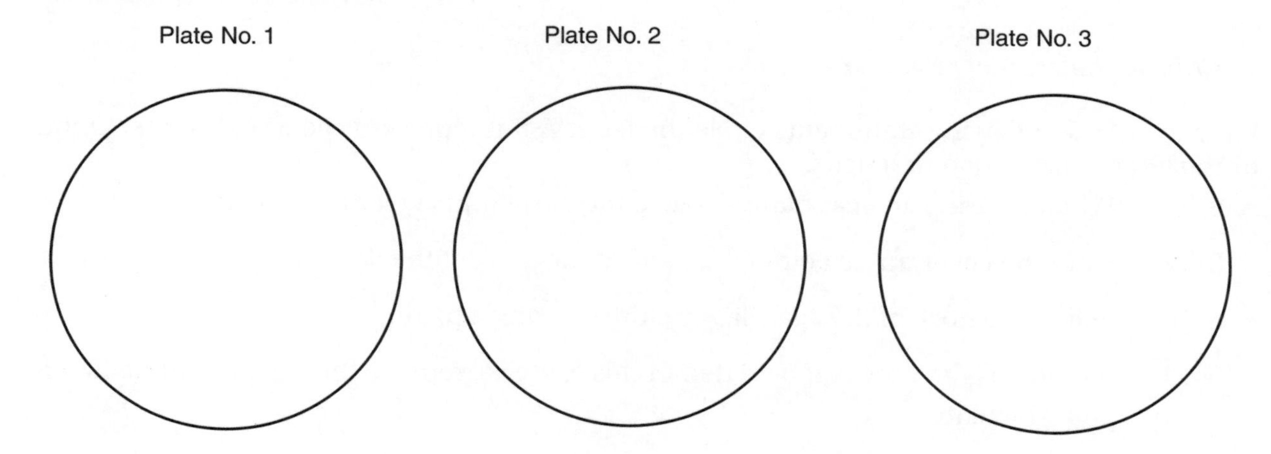

Questions

A. Completion

1. A freshly poured plate should be rotated carefully in a ________________ .
2. What is the predominant shape (i.e., form) of subsurface colonies?
   ____
3. Which method of separating organisms, streak or pour plating, seems to achieve the best separation? __________________

4–5. Give 2 reasons why the nutrient agar must be cooled to 45°C before inoculating and pouring.

4. ____
5. ____

6. The name of the instrument used to help count colonies on plates is the
   __________________ __________________ __________________ .
7. This technique is said to be roughly quantitative because the loop used contains only approximately ________ ml.
8. If your tubes solidify after inoculation but before you pour them into plates, what happens to the bacteria in the tubes when you now remelt the tubes?
   ____

9. Give one reason why you should avoid slopping agar up on the cover while mixing the sample.

____ .

10. When you incubate these plates, the cover should be __________ .

B. Choose the correct response

For each of the following statements circle the letter A if it represents good aseptic technique or the letter B if it is bad practice.

A B 11. When pouring an agar plate, remove the cover and lay it on the bench.

A B 12. Hold the cover up just enough to admit the neck of the flask.

A B 13. Flame the neck of the agar flask before pouring a plate.

A B 14. As soon as the tube is inoculated in this exercise, remove the cap and immediately pour the plate.

A B 15. Heat the lip of the tube so that the agar sizzles as it is being poured into the plate.

A B 16. Incubate your plates with the cover up.

A B 17. After streaking a plate with a pure culture, lay the loop down on the bench.

A B 18. Carefully remove the cap on a tube, and without setting it down, make your transfer.

A B 19. Flame your inoculating loop before and after making a transfer.

A B 20. While attempting to remove a tube cap, your loop accidentally touches the sleeve of your lab coat.

Name ______________________________ Date ________ Grade ________

10. Quantitative Dilution and Spectrophotometry

Results and Observations

A. Determination of Bacterial Count Per Unit Volume

1. Count ____________
2. Dilution counted ______
3. Bacteria ____________ per ml

B. Optical Density vs. Bacterial Count

Dilution	*Absorbance*	*Bacteria per ml*	
undiluted	________	________	(same as line A.3)
1:2	________	________	
1:4	________	________	
1:8	________	________	
1:16	________	________	

Absorbance

1.0

0.8

0.6

0.4

0.2

1/16 1/8 1/4 1/2 none Dilution

Bacteria/ml

___ ___ ___ ___ ___ x 10(add exponent)

Questions

1. How many pipets are you allowed to remove from the container and place on the bench top while working? _____

2–4. List the only three places where pipets should be seen in a laboratory.

2. ______________________________
3. ______________________________
4. ______________________________

5. A 1:2 dilution of a culture has ________ as many bacteria per ml as the culture itself.

6. A unit volume of a sample is usually one _________ .

7–8. Bacterial numbers are directly proportional to ______________________ and inversely proportional to ______________________ .

9. The dilution factor used to multiply a plate count by to obtain the number of bacteria per unit volume is the ____________________ of the dilution of the plate on which 30–300 colonies are found.

10. The preferred way to report 5,365,000 bacteria per ml is ____________ .

11. Duplicate plates were counted with an average of 189 colonies present at the 10^{-3} dilution. The bacteria per ml of original sample is ________________ .

12. No plate at any dilution had 30–300 colonies but the 10^{-1} dilution had 25. The bacterial count per ml is __________________ .

13. A culture has 2×10^9 bacteria per ml. The 1:4 dilution has _______________ bacteria per ml.

14. If a plate has a ____________________ over half or more of the plate, it should not be used for counting.

15. The technique of making dilutions to count bacteria is called the ____________________ ____________________ method.

16. For purposes of making dilutions, one ml of water weighs ____ g.

17. In order to be a countable plate, it should have between __________ colonies.

18–20. From **your data graph of absorbance vs. bacteria** per ml on this report form, what would the following absorbance readings give as bacteria per ml?

absorbance	Bacteria per ml
0.8	18. ____________
0.5	19. ____________
0.1	20. ____________

Name ______________________________ Date ________ Grade ________

11. Smear Preparation and Simple Staining

Results and Observations

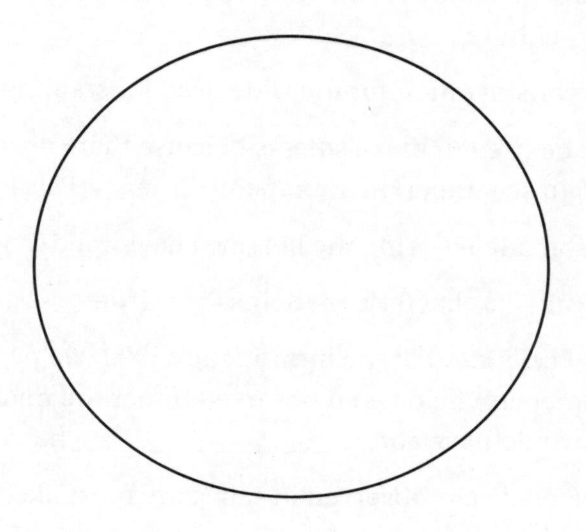

Organism Smear from teeth

Morphology ______________________

Magnification ______________________

Stain ______________________

Questions

1. Bacterial cells are difficult to see because their index of refraction is very similar to _________ .
2. A good smear should appear only slightly _________________ .
3. A slant culture is generally considered a better source of organisms for staining than is a ____________ culture.
4. An important consideration for the slide used for staining is that it be ___________ .
5. Tap water can be used to make smears because there are too few bacteria (about 10,000 per ml) to see under oil immersion. True ___ False ___
6. A smear is best made if during the heating stage (mild/high) heat is used.
7. Virtually everyone has bacteria in their saliva. True ___ False ___
8. The number of cells cited in the introduction (500,000 per ml) will result in about 1 cell per oil immersion field when observed under the microscope. If you had ten cells per field, you would have about __________________ bacteria per ml of your saliva.
9. The large nucleated cells observed in saliva are most likely __________________ .
10. If cells stained with methylene blue are washed too long, the dye will be ______________________ .

Name ______________________________ Date ________ Grade ________

12. Gram Stain

Results and Observations

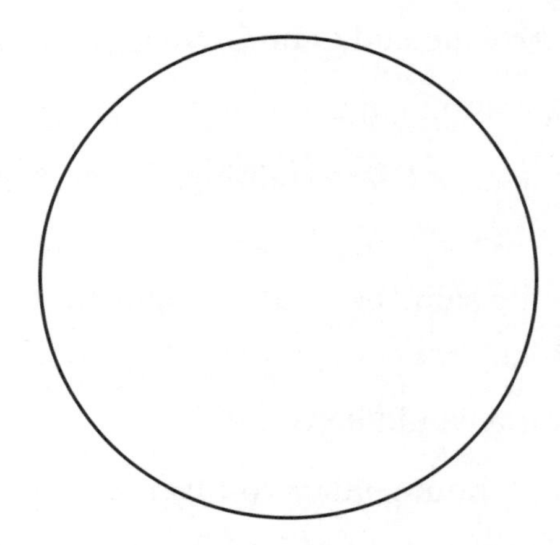

Organism *Escherichia coli*

Morphology ____________________

Gram Reaction ____________________

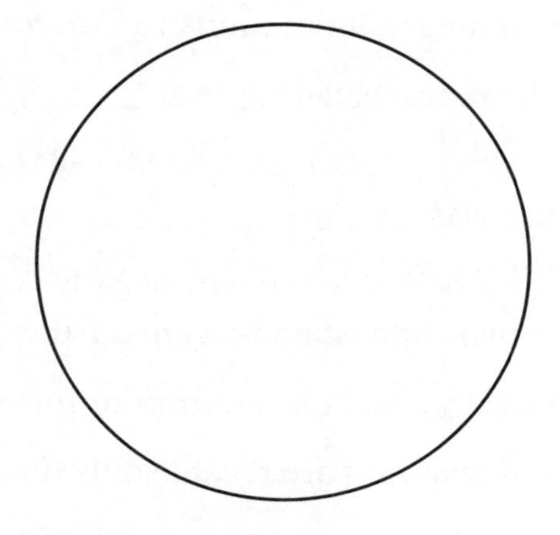

Organism *Staphylococcus aureus*

Morphology ____________________

Gram Reaction ____________________

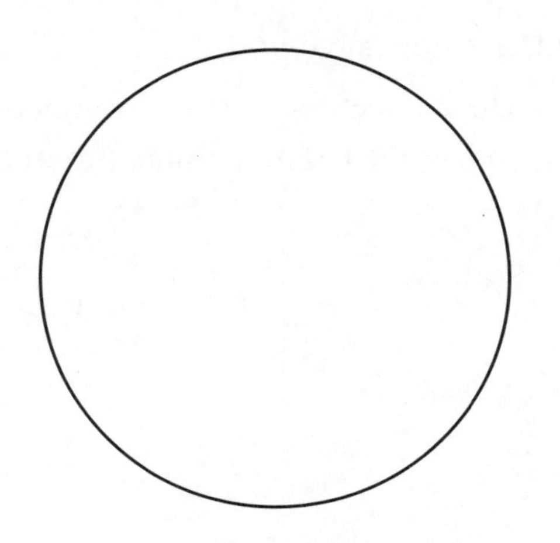

Organism *Bacillus cereus*

Morphology ____________________

Gram Reaction ____________________

Questions

1. A preparation made by mixing a loopful of water and a bit of agar slant culture on a glass slide is called a ___________ .
2. The Gram stain is an example of a ___________________ (type) stain.
3. All members of the genus *Escherichia* would be expected to be Gram-_____________ .
4. Endospores usually appear ____________ in a completed Gram stain.
5. If your *Staphylococcus aureus* cells appear pink after Gram staining, the most likely cause was ___________ ______________________ .
6. Gram-positive and Gram-negative cells have the same color after application of the first two Gram stain reagents. True____ False____
7. Generally speaking, endospore-forming bacteria would be Gram-_________________ .
8. Gram reactions are reliable only for cultures ___ hours old or younger.
9. Unless the procedure specifically indicates otherwise, Gram stain smears are **first** __________ ___________ .
10. A smear when properly made should look (like milk/very dense/faintly cloudy).
11. The most critical step in making the Gram stain is the application of the ______________________ .
12. A culture that is too old may appear Gram-______________ when stained.

13–28. Fill in the table below to show the changes in Gram-positive and Gram-negative bacteria during each major step of the Gram-staining process.

Reagent	Purpose	Microscopic Appearance of	
		Gram + Organisms	Gram – Organisms
13–16.			
17–20.			
21–24.			
25–28.			

Name ________________________________ Date __________ Grade __________

13. Acid-Fast Stain

Results and Observations

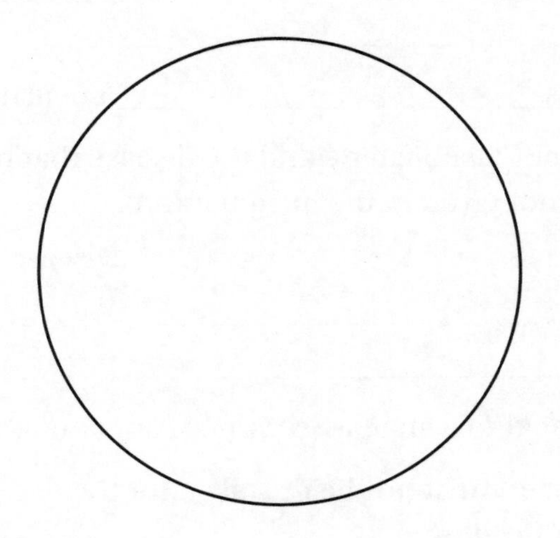

Staphylococcus aureus
+
Organisms *Mycobacterium smegmatis*

Magnification ______________________

Stain ______________________

Questions

1. Carbolfuchsin stain is prepared by adding ____________ to basic fuchsin.

2–4. The acid-fast stain identifies microorganisms with a high ______________ content. The primary stain is not washed out by the ________________ decolorizer, and the cells are said to be _________________ .

5. Members of the genus _________________________ are usually all acid-fast.

6–9. Name 2 pathogenic acid-fast bacteria and the disease that each one causes (consult a textbook for answers not given in the introduction).

Organism	Disease
6. ____________________ ____________________	7. ______________________
8. ____________________ ____________________	9. ______________________

10. Among bacteria, the acid-fast state is (common/uncommon).

11. The acid-fast decolorizer **must not** be mistaken for the decolorizer in the ____________ stain procedure.

12. The primary stain in the acid-fast procedure is _________________________ .

13. When placing mixed cultures on a slide, it is important to ___________ the inoculating loop between cultures.

14–25. Fill in the table below to show the changes in acid-fast and non-acid-fast bacteria during each major step of the Ziehl-Neelsen acid-fast method.

Reagent	Purpose	Appearance of Cells	
		Acid-Fast	Non-Acid-Fast
14–17.			
18–21.			
22–25.			

Name ______________________________ Date __________ Grade __________

14. Endospore Stain

Results and Observations

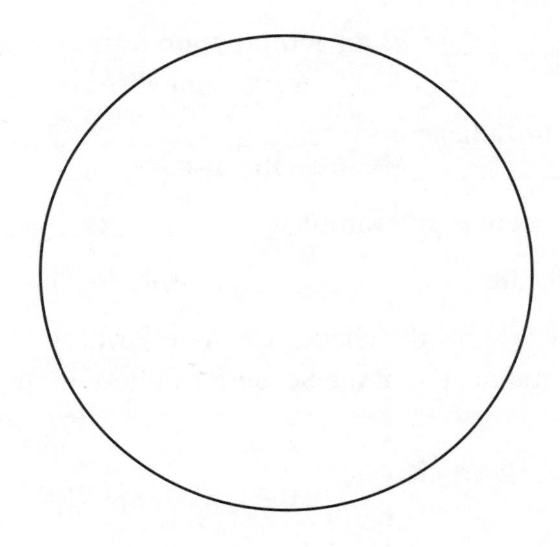

Staphylococcus aureus

+

Organisms *Bacillus cereus*

Position of Spore ______________________________

Magnification ______________________________

Stain ______________________________

Questions

1. In the Schaeffer-Fulton endospore staining method, ______________ is the decolorizer.
2. The endospore is (more/less) resistant to environmental stresses than the vegetative cell.
3. In bacteria, endospore formation is usually triggered by ____________ changes in the environment.

4. The primary stain in the Schaeffer-Fulton method is ______________ ______________ .

5. A trace element often used in media to stimulate endospore formation is ____________________ .

6. Individual species of endospore forming bacteria characteristically form spores that have specific __________ , ___________ , and may swell the cell or not.

7–10. The genus __________________ is an aerobic sporeforming genus. One species, ________________ ________________ , causes the disease ___________________ .

11–14. The anaerobic sporeforming genus is __________________ . One species, _______________ ____________ , causes the disease ________________ .

15. The Schaeffer-Fulton stain is an example of a ________________ (type) stain.

16. Sporeforming bacteria are ______________ (morphology).

17–28. Fill in the table below to show the changes in sporeforming and non-sporeforming bacteria during each major step of the Schaeffer-Fulton method.

Reagent	Purpose	Appearance of	
		Spores	Vegetative Cells
17–20.			
21–24.			
25–28.			

29–31. Identify the location of the endospore in each of the following drawings.

Location

29.______________________________

30. ______________________________

31. ______________________________

Name ____________________ Date ________ Grade ________

15. Starch Hydrolysis

Results and Observations

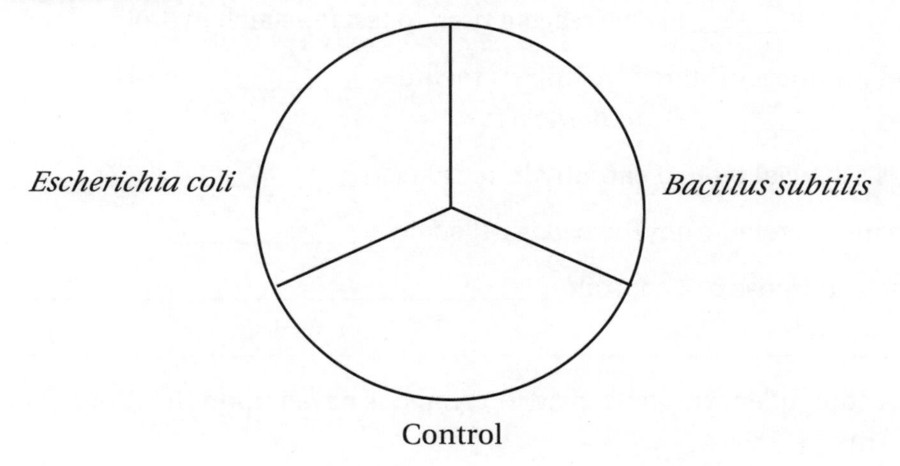

Organism	Starch Hydrolysis	
	+	−
Control		
Bacillus subtilis		
Escherichia coli		

Questions

1. A positive test for starch hydrolysis appears as ______ ___________ or __________ .
2. ____________________ is the general name of the enzyme that hydrolyzes starch.
3. ____________________ is the reagent used to test for starch hydrolysis.

4–5. The end products of starch hydrolysis include _____________ and ____________________ , to name a few.

6. Starch is a special type of carbohydrate called a ________________________ .
7. An enzyme excreted from the cell is called an ________________________ .
8. What is the purpose of a control? ____

   ____
9. The detection of enzymatic activity on starch is based upon the disappearance of the starch. True ____ False ____
10. The specific amylase, glucoamylase, results in ________________ as the end product.

Name ____ Date _________ Grade __________

16. Lipid Hydrolysis

Results and Observations

Organism	Lipid Hydrolysis	
	+	–
Control		
Staphylococcus aureus		
Proteus mirabilis		

Questions

1. Fats and oils are hydrolyzed by enzymes generally called ________________ .

2–3. _________________ and ___________ _____________ are the immediate end products of lipid hydrolysis.

4. After incubation, lipid hydrolysis is tested for by the addition of a reagent.
True ___ False ___

5. The color of a positive test for lipid hydrolysis is _________ __________ .

6. __________________ is another term for lipid hydrolysis.

7. ___________ ___________ is a dye which serves as the indicator of lipolytic activity.

8–9. A lipase produced by *Clostridium perfringens* causes ________________ of red blood cells by removing ___________________ from certain membrane lipids.

10. An important property of lipids in the presence of water, is that they are
_______________________ .

Name ______________________ Date ________ Grade ________

17. Protein Hydrolysis (Gelatin and Casein)

Results and Observations

A. Drawing of Overlay Plate (do only one)

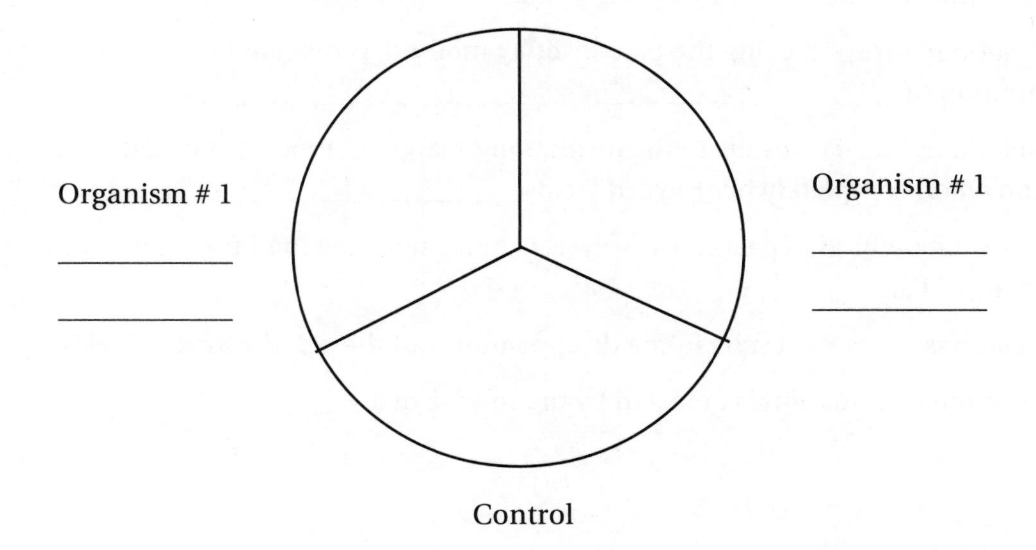

B. Combined Table

Organism	Gelatin Hydrolysis		Casein Hydrolysis	
	+	–	+	–
Control				
Serratia marcescens			XXXX	XXXX
Escherichia coli				
Bacillus cereus	XXXX	XXXX		

Questions

Gelatin

1. At room temperature, gelatin is a ____________ .

2. Gelatin is derived from the animal substance called ________________ .

3. The collective name for the enzymes that hydrolyze gelatin is __________________ .

4. Gelatin hydrolysis, using the plate overlay method, is detected by the addition of a solution of ______________ ______________ ______________ .

5–6. Gelatin hydrolysis results in the formation of larger compounds called ____________ and when completely hydrolyzed yields _________ _________ .

7. The tube method of gelatin hydrolysis is more sensitive than the plate overlay method. True ___ False ___

8. Gelatinase activity results in the disappearance of the gelatin. True ___ False ___

9. Collagen degradation is detected by the release of a ___________ .

Casein

10–11. The immediate end products of casein hydrolysis include ______________ and ____________ ____________ .

12. Casein is a ______________ found in milk.

13. The group of enzymes that can hydrolyze casein are collectively called ________________ .

14. The term used to describe casein hydrolysis specifically is ______________________ .

15. The colloidal nature of milk serves as the basis for interpreting casein hydrolysis. True ___ False ___

16. A special chemical reagent must be added to test for casein hydrolysis. True ___ False ___

Name ______________________________ Date ________ Grade ________

18. Litmus Milk Reactions

Results and Observations

Enter codes (at bottom of table) in the boxes provided for observations.

Organism	Type of Reaction		
	24 hr	48 hr	7 d
Control			
Acinetobacter calcoaceticus			
Enterococcus faecalis			
Escherichia coli			
Pseudomonas aeruginosa			

Reaction codes to be used:

NC = no change
A = acid
AC = acid curd
W = whey
G = gas
P = peptonization
R = reduction
RC = rennet curd
ALK = alkaline

Questions

1. Peptonization is often accompanied by the production of ________________.

2. Peptonization in milk results from the hydrolysis of ____________.

3. A rennet curd is usually observed only under ____________________ conditions.

4. The reduction of litmus results when ______________ are transferred to the molecule.

5. When gas production is so vigorous that the acid curd is blown to shreds, it is called __________ ________________.

6. If the lactose in litmus milk is fermented by an organism, then the color of the litmus will be ______.

7. When litmus turns white, it indicates that ________________ has taken place.

8–9. In litmus milk, the litmus serves as both a ________________ _______ and a _______ ________________.

10. ______________ is the main carbohydrate in milk.

11. When milk protein is hydrolyzed in litmus milk, it is called ___________________.

12. When a curd develops in litmus milk, how could you determine whether it was a rennet or acid curd?

    ____

    ____

13. A ______________ color is the normal color of litmus milk.

Name ______________________________ Date ________ Grade ________

19. Nuclease Activity

Results and Observations

Draw the appearance of the overlay plate.

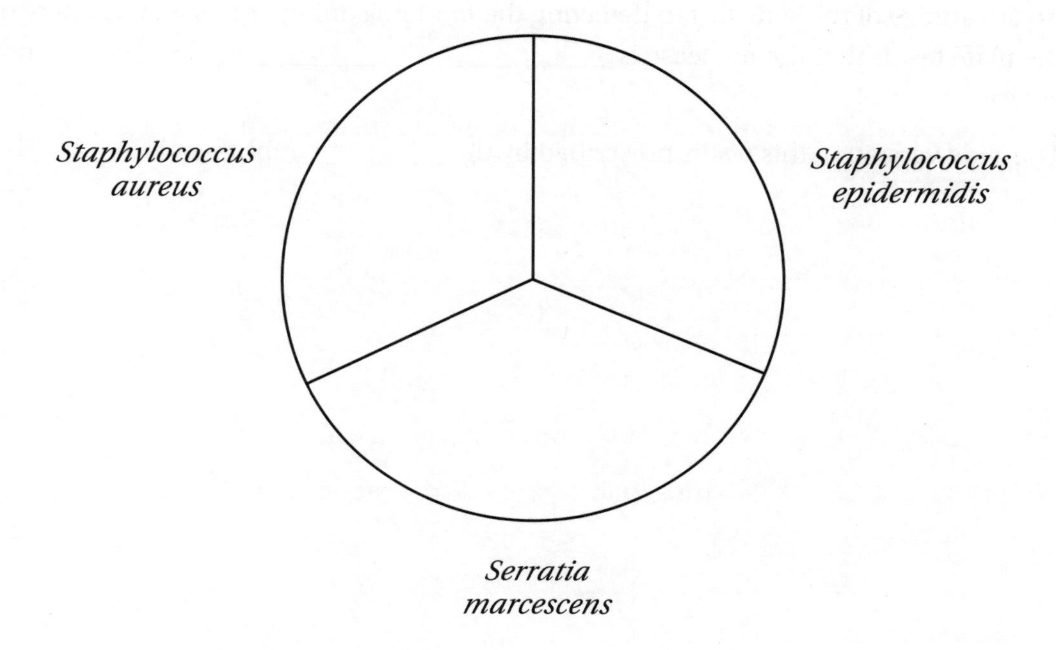

Organism	DNA Hydrolysis	
	+	−
Serratia marcescens		
Staphylococcus aureus		
Staphylococcus epidermidis		

Questions

1. To demonstrate DNA hydrolysis, ______________ ___________ dye is used in the medium.
2. To precipitate DNA in this medium (without dye), ___ ________ is used.
3. The organism used in this exercise which causes food poisoning is ____________________ ____________ .
4. An advantage of the slide test in detecting the food poisoning organism used here over the plate test is that the nuclease is ________________________ and live cells are not necessary.
5. The slide version of this test is prescribed by the _________ (abbreviation).

Name ______________________________ Date ________ Grade ________

20. Hemolysis

Results and Observations

Make a drawing of the plate.

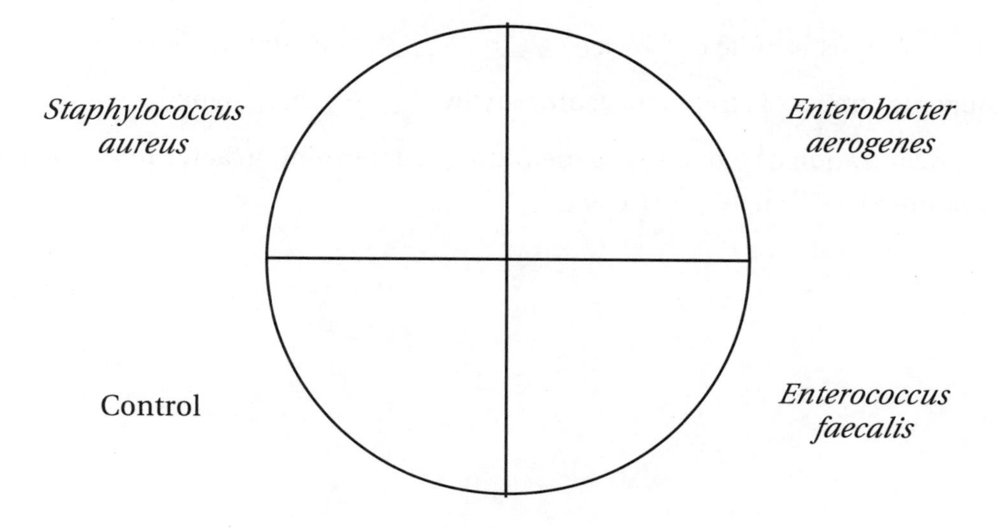

Organism	Type of Hemolysis		
	alpha α	beta β	non-hemolytic
Enterobacter aerogenes			
Enterococcus faecalis			
Staphylococcus aureus			

Questions

1. Beta hemolysis is indicated by a __________ zone around a colony.
2. Hemolysis is best observed against a _________ background.
3. The medium used to detect hemolysis production is ___ ________ ________ _______ .
4. Hemolysins are (exoenzymes/endoenzymes).
5. Alpha hemolysis is indicated by a ___________ zone around a colony.
6. Pathogenic strains of streptococci are usually _______ hemolytic.
7. The concentration of blood in the medium is a determining factor for the length of incubation at 37°C. True ___ False ___

Name ______________________________ Date ________ Grade ________

21. Sugar Fermentations

Results and Observations

Record your results using the codes given at the bottom of the table.

Organism	Time	Carbohydrates		
		PR-glucose	PR-sucrose	PR-lactose
Control	24 hr			
	48 hr			
	72 hr			
	7 d			
Alcaligenes faecalis	24 hr			
	48 hr			
	72 hr			
	7 d			
Bacillus subtilis	24 hr			
	48 hr			
	72 hr			
	7 d			
Escherichia coli	24 hr			
	48 hr			
	72 hr			
	7 d			
Staphylococcus aureus	24 hr			
	48 hr			
	72 hr			
	7 d			

Recording codes: NC = no change, AG = acid and gas, A = acid, ALK = alkaline

Questions

1–2. Amino acid fermentation leads to the production of ________ ________ ________ and ________________ .

3. The specific enzymes causing the fermentation of a sugar belongs to the _______ enzyme type (use prefix).

4. If phenol red turns a magenta color, it means ____________ has been produced.

5–6. The small inverted tube found within a larger tube is called a __________ ____________ tube and is used to trap ________ .

7. When phenol red changes to yellow, this indicates that _________ has been produced.

8–9. The two pH indicators most often used in a microbiology laboratory are ____________ ____________ and ____________ ________________ .

10–11. The process of fermentation involves two groups of enzymes known as ________________________ and _______________ ______________.

12–15. List the 4 essential parts or ingredients of a fermentation tube.

12. ____________________________ 14. __________________________

13. ____________________________ 15. __________________________

16–19. List four fermentable carbohydrates other than those used in this exercise. (Refer to a microbiology text or Exercise 33 for help.)

16. ________________________ 18. ____________________

17. ________________________ 19. ____________________

20. ________________________ is the term that refers to the anaerobic breakdown of carbohydrates.

21–24. What is the pH of the acid color change of the indicator used in this exercise? What colors denote a neutral, acid, or alkaline reaction?

21. pH of acid color change _______

22. neutral _______

23. acid _______

24. alkaline _______

25. If a carbohydrate broth does not change color after it has been inoculated and incubated, how can you tell whether the unchanged color is due to failure of the organism to grow or failure to ferment the carbohydrate? ________________________________

____

Name ____________________ Date ________ Grade ________

22. Hydrogen Sulfide Production

Results and Observations

Enter data in the table using codes found at the bottom.

Organism	Fermentation Reactions		H_2S Production	
	Glucose	Lactose	+	–
Control				
Escherichia coli				
Proteus vulgaris				

NC = no change
A = acid
AG = acid and gas

Questions

1. When hydrogen sulfide is released by microorganisms, it is in the form of a ________ .
2. The production of hydrogen sulfide is an (aerobic/anaerobic) reaction.
3. Under natural conditions, where might you expect to find hydrogen sulfide production by microorganisms? ______________________________
4. ______________ is the color of a positive test for hydrogen sulfide.
5. In order to produce hydrogen sulfide, microorganisms must possess the enzymes to attack ____________-containing compounds.
6. The formation of the compound __________ __________ provides the basis for the detection of hydrogen sulfide.

7–10. List four media used to detect hydrogen sulfide production.

7. ______________________________ 9. ______________________________

8. ______________________________ 10. ______________________________

11. ____________ ____________ is the salt added to the medium to detect hydrogen sulfide production.
12. ______________ _______ is the pH indicator used in KIA medium.

13–14. The acid colors of carbohydrate fermentation reactions must be read at ____ hours and at ____°C to be valid.

15–16. __________ and ________ are considered to be multipurpose types of media.

17. In the absence of sugars, does protein decomposition result in a shift toward the acid or alkaline side? Explain. ______________________________

____

Name ____________________ Date ________ Grade ________

23. IMViC Reactions

Results and Observations

Enter results in the table using the key symbols at the bottom.

Organism	IMViC Reactions			
	I	M	Vi	C
Control				
Escherichia coli				
Enterobacter aerogenes				

+ = positive test
– = negative test

Questions

1. _______________ is a breakdown product of tryptophan.

2. The *Enterobacter-Klebsiella* are referred to as ________ ________ fermenters.

3. The sugar used as the basis for the methyl red test is _____________ .

4–5. Name two media that can be used to detect citrate utilization.

4. ___________________________

5. ___________________________

6. __________________ is the medium used to test for indole production.

7. ___________________ is the chemical reagent used in the indole test.

8. The pH must be below _____ for a positive methyl red test to occur.

9–10. Name the two major products produced by Voges-Proskauer positive organisms.

9. _________________________________

10. _________________________________

11–12. ______________________ and ______________________are the reagents used in the Voges-Proskauer test.

13. The members of the *Enterobacteriaceae* or the enteric bacilli for which the IMViC tests are most useful belong to the ________________ group.

14. Those enterics that do not fall into the two distinctive categories formed by *Escherichia coli* and *Enterobacter aerogenes* are collectively referred to as _________________________ .

15–22. Indicate the color of positive and negative tests for each of the following:

Reaction	***Positive Test***	***Negative Test***
Indole	15. __________	16. __________
Methyl red	17. __________	18. __________
Voges-Proskauer	19. __________	20. __________
Citrate	21. __________	22. __________

Name ______________________________ Date ________ Grade ________

24. Catalase and Oxidase Production

Results and Observations

Organism	Catalase Production		Oxidase Production	
	+	−	+	−
Control				
Staphylococcus aureus			XXXX	XXXX
Enterococcus faecalis			XXXX	XXXX
Pseudomonas aeruginosa	XXXX	XXXX		
Escherichia coli	XXXX	XXXX		

Questions

A. Catalase

1–2. _______________ is a respiratory enzyme found in most aerobic organisms. It is tested for by the addition of the chemical reagent __________________ ____________________ .

3–6. The catalase test is especially valuable in distinguishing between the Gram-positive cocci ______________________________ (catalase +) and ________________________________ (catalase –), and in distinguishing between the Gram-positive long rods _____________________ (catalase +) and _________________ (catalase –).

7. _______________ organisms do not produce catalase and are poisoned by the accumulation of hydrogen peroxide.

B. Oxidase

8. An obligate anaerobe would be expected to give a _____________ test for oxidase.

9. A __________________________________ color appears in a positive oxidase test.

10. The chemical _______________________________ is added to test for the production of oxidase.

11. A positive oxidase test depends upon the presence or absence of ____________________ ____ in the cell.

12. Oxidase production has no value in classification. True ___ False ___

Name ______________________________ Date ________ Grade ________

25. Nitrate Respiration

Organism	Nitrite Test			Results of Zinc Test *if* Nitrite Test Negative	Conclusion from Results
	Color	+	–		
Control					
Bacillus subtilis					
Pseudomonas aeruginosa					
Staphylococcus aureus					
Soil sample					

Questions

1. A red color indicates a __________ test for nitrite when the correct reagents are added.

2–3. In addition to microbial enzymes, nitrates can be reduced to nitrites chemically by the addition of powdered __________ and __________ .

4. If nitrite is absent and zinc and acid are then added, the development of a red color is a ______________ test for the presence of nitrate.

5–6. List the reagents used to test for nitrite.

5. ______________

6. ______________

7–9. Give the reaction color for the following when sulfanilic acid and dimethyl-1-naphthylamine reagents are added.

Reaction	*Reaction Color*
NO_3^- —> NO_2^-	7. ____________
NO_3^-	8. ____________
NO_3^- —> NH_3	9. ____________

10. An enzyme that reduces nitrate to nitrite is ______________ ______________.

11. An enzyme that reduces nitrite is ______________ ______________.

12. Reduction of nitrate to nitrite is favored by (aerobic/anaerobic) conditions.

13. It is preferable to perform this test by periodic testing of some of the culture over a period of several days. Why? ______________________________

14–16. Explain each of the following:

14. Nitrite test negative, no nitrate present ______________________________

15. Nitrate test negative, nitrite present ______________________________

16. Nitrite test negative, nitrate present ______________________________

17. Briefly explain why a carbohydrate such as glucose should not be added to this medium for denitrification determination. ______________________________

____

Name ______________________________ Date ________ Grade ________

26. Ammonification

Results and Observations

Enter results in the Table using the code found below.

Organism	Ammonia Production		
	2 d	4 d	7 d
Control			
Bacillus subtilis			
Pseudomonas aeruginosa			
Soil Sample			

+ = ammonia produced

– = no ammonia produced

Questions

1. The test for the production of ammonia uses ______________ reagent.
2. A positive test for ammonia using the reagent in question 1 appears ____________ .
3. Cultures should be incubated over a period of several days. Why? ______________
   ____
4. Why is soil used as an inoculum? ______________________________________
   ____
5. The process which involves the release of ammonia is called ____________________ .
6. Ammonia is produced when an organism utilizes ________________ for its carbon.
7. When carbon is in excess, ammonia will be __________________ .

Name ______________________________ Date ________ Grade ________

27. Urea Hydrolysis

Results and Observations

Organism	Urea Hydrolysis	
	+	–
Control		
Escherichia coli		
Proteus vulgaris		

Questions

1–3. The enzyme ______________ splits urea into ______________ and ______________ ______________ .

4. A ________ color appears in urea broth as a result of hydrolysis.

5. ______________ ________ is the pH indicator used in urea broth.

6. The pH of urea broth must be at least ______ in order to give a positive urea hydrolysis test.

7. The genus of bacteria used in this exercise which hydrolyzes urea is ________________.

8–9. The substance that gives the positive result is ______________ and does so by causing an increase in ______ .

Name ____________________ Date ________ Grade ________

28. Temperature

Results and Observations

Table of Growth Response

Enter growth as + or – at the upper left of the diagonal for 2 days' incubation and the lower right for 7 days' incubation.

Organism	5°C	20°C	30°C	40°C	50°C	60°C	Growth[a] Temperature		
	2d 7d	2d 7d	2d 7d	2d 7d	2d 7d	2d 7d	Max	Min	Opt
B. stearothermophilus									
Escherichia coli									
Pseudomonas fluorescens									
Vibrio marinus									

[a]If these can be determined from your data, they are approximate only.

This type of recording of growth does not show it very clearly, but was there a difference in amount of growth between 2 and 7 days? If so, which bacteria?

Questions

1–5. Give the name of the organism temperature grouping for each of the following:

1. Grows at 0°C ______________________________
2. Optimum growth at 15°C ______________________________
3. Grows at 70°C ______________________________
4. Optimum growth at 55°C but grows at 37°C ______________________________
5. Grows best at 37°C ______________________________

6. Growth could be better quantified by using a spectrophotometer and measuring ______________________________.

7. The optimum growth temperature (is/is not) necessarily the optimum temperature for all other cellular activities.

8. All strains of the species *Pseudomonas fluorescens* (have/do not have) the same optimum temperature.

9. The temperature above which an organism will not grow is said to be the ____________________ temperature.

10. Growth below 0°C would require, among other things, a ______________ water phase.

11. A ______________________ grows optimally between 20°C and 45°C.

12. Growth may or may not be signified by a ______________ of cells on the bottom of a culture tube.

Name ______________________________ Date ________ Grade ________

29. Osmotic Pressure

Results and Observations

Table of growth on NaCl agar

Organism	0.5%		5%		10%		15%		20%	
	48 hr	1 wk	48 hr	1 wk	48 hr	1 wk	48 hr	1 wk	48 hr	1 wk
Escherichia coli										
Halobacterium salinarium										
Staphylococcus aureus										
Saccharomyces cerevisiae										

Comments:

Questions

1. An organism requiring salt to grow is called a ____________________.

2. When the organism in question 1 grows, the solution on which it grows would be ____________________ (or nearly so) to the cell cytoplasm.

3. If the organism grown in question 2 is suddenly placed in fresh water it would __________.

4–5. In the opposite of question 3, when a cell _________ water to the outside, it is called ____________________.

6–7. Movement of water from one side of a membrane to the other depends on ______ water, the amount of which in turn depends on the amount and kind of ____________ present.

8. The species of archaebacteria studied in this exercise is ____________________ ________.

9–10. An organism growing in syrup would be called an ____________________ and would most likely be a _________________ (organism group).

11–12. An organism growing on a dehydrated food would be called a ___________________ and would most likely be a _________________ (organism group).

13. The amount of free water in a solution is sometimes referred to as _______________ ______________.

14. Where in nature would you go to find a halobacterium? ________________________ ____

Name ______________________________ Date ________ Grade ________

30. pH

Results and Observations

Table of pH results

Organism		pH						
		4.0	5.0	6.0	7.0	8.0	9.0	10.0
Bacillus sp.	48 hr							
	7 d							
Saccharomyces cerevisiae	48 hr							
	7 d							
Staphylococcus aureus	48 hr							
	7 d							
Escherichia coli	48 hr							
	7 d							
Enterococcus faecalis	48 hr							
	7 d							

Questions

1–2. Organisms growing at high pH are generally called ____________________, and those growing at low pH are called ____________________.

3. A genus of archaebacteria with a very low optimum growth pH and a very high optimum growth temperature is ____________________.

4. A genus of bacteria found in acid mine wastes is ____________________.

5–6. Name a mineral from which the acid is produced in acid mine waters and the acid produced (consult a textbook).

5. Mineral ____________________

6. Acid ____________________

Name ______________________________ Date ________ Grade ________

31. Ultraviolet Light

Results and Observations

Draw your results and those of the other two lamps.

Organism ______________________________

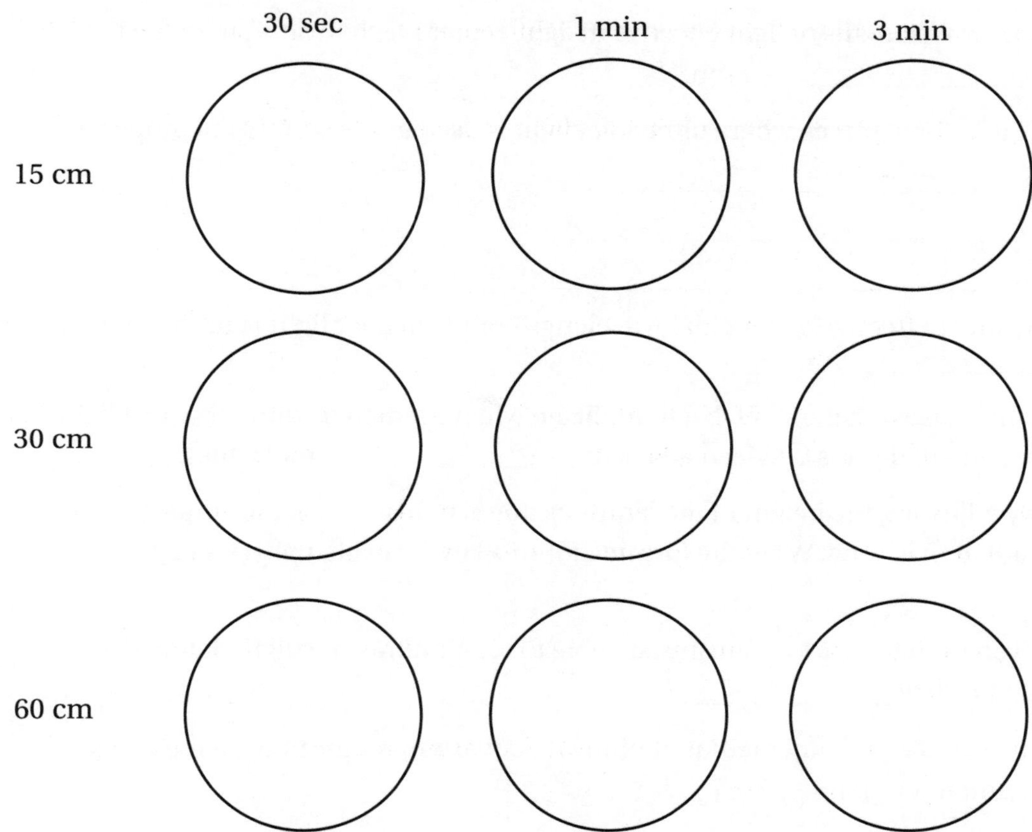

Comments and observations on relative colony numbers on the plates:

Questions

1–3. Name three factors that limit the use of ultraviolet light as a sterilizing agent.

1. ____
2. ____
3. ____

4. Ultraviolet light primarily affects the cell's __________ .

5. The wavelengths of light effective in light-repair mechanisms are in the range of ____________________ nm.

6–8. Name three places where ultraviolet light is used for its germicidal properties.

6. ________________________________
7. ________________________________
8. ________________________________

9. The most effective germicidal wavelength of ultraviolet light is in the neighborhood of ______________ nm.

10. One cause of failure of DNA to replicate when irradiated with ultraviolet light, is the formation of dimers between adjacent _________________ molecules.

11. By adjusting the height of the lamps as done in this exercise, a simple physical law of radiation is used. Write the formula for this law (consult a physics text).

12. Even though some organisms are able to repair ultraviolet light damage, surviving cells often carry ____________________.

13. In addition to light-mediated photoreactivation, repair may take place as __________-mediated repair.

14. Define photoreactivation:

Name ______________________________ Date ________ Grade ________

32. Bacterial Motility

Results and Observations

A. Wet Mount Preparation

Organism	Motility		Description of Motility
	+	−	
Bacillus subtilis			
Pseudomonas aeruginosa			
Staphylococcus aureus			

B. Soft Agar Method

Organism	Motility		Extent of Growth	Amount of Growth
	+	−		
Bacillus subtilis				
Pseudomonas aeruginosa				
Staphylococcus aureus				

Questions

1. The _________ _________ is the more sensitive of the motility study methods.
2. The number and location of flagella can be utilized for _______________ purposes.
3. A ___________ organism migrates through the agar gel in a motility medium.
4. Prokaryotic flagella usually can (be seen, not be seen) on live cells.
5. Name a eukaryotic organism group showing gliding motility: __________________
6. Bacteria as a group may show the following (cross out those **not** applying): gliding/ flagellar motility/ amoeboid movement/ Brownian movement/ cilia.
7. Brownian movement results when a cell is bombarded by (bacteria/ water molecules/ eukaryotes).
8. A necessary requirement for the soft agar method is a small concentration of ___________ .
9. The ability of an organism to move toward a nutrient or away from harm is called ___________________ .
10. Name a group of organisms with cilia: __________________

Name ____ Date __________ Grade __________

33. Identification of an Unknown Organism

Data Collection Sheet

Unknown No. ___________

Character No. **Character Scores: ND = Not Done, Missing**

Cell Morphology and Staining

1. Shape: 0- ND; 2- straight rod, or coccobacillus; 3- curved or spiral rod; 5- spirochete; 6- coccus; 7- branched
2. Length of rod or diameter of coccus: 0- ND; 2- <0.5 μm; 3- 0.5 to 3.0 μm; 4- >3.0 μm
3. Width: 0- ND; 1- coccus; 2- <1 μm; 3- >1 μm
4. End of cell: 0- ND; 1- coccus; 2- round end
5. Cell arrangement: 0- ND; 2- more than 50% single; 3- more than 50% pairs; 4- chains <5 cells per chain; 5- chains >5 cells per chain; 6- packets (4 or 8) or irregular clusters of cocci; 7- branching filaments
6. Endospore: 0- ND; 1- none; 2- cell swollen terminal or subterminal; 3- swollen central; 4- cell normal terminal or subterminal; 5- normal central
7. Air relation: 0- ND; 2- aerobic or facultatively anaerobic; 3- obligate anaerobe; 4- microaerophilic
8. Gram reaction: 0- ND; 2- positive or variable; 3- negative
9. Motile: 0- ND; 1- non-motile; 2- motile
10. Acid fast: 0- ND; 1- no; 2- yes

Growth Characteristics

11. Colony size (48–72 hours well isolated): 0- ND; 2- <1 mm; 3- 1-5 mm; 4- >5 mm
12. Colony elevation (48–72 hours well isolated): 0- ND; 2- flat (effuse); 3- raised; 4- umbonate; 5- convex; 6- pulvinate
13. Colony margin (48–72 hours well isolated): 0- ND; 1- entire; 2- undulate; 3- lobate; 4- curled; 5- erose; 6- filamentous; 7- spreading or swarming over whole plate
14. Colony color: 0- ND; 1- non-pigmented, off-white, white, or gray-white; 2- blue; 3- violet or purple; 4- brown or black; 5- red or pink; 7- orange; 8- green or yellow-green; 9- yellow
15. Pigment solubility: 0- ND; 1- confined to colony (score white, off white and non-pigmented here); 3- diffusible (in agar rather than colony)
16. Pigment fluorescence: 0- ND; 1- not fluorescent; 2- fluorescent (ultraviolet)
17. Colony density (48–72 hours well isolated): 0- ND; 2- transparent (read print through center) or translucent; 4- opaque
18. Turbidity in broth: 0- ND; 1- none; 2- slight; 3- moderate to heavy
19. Sediment in broth: 0- ND; 1- none; 2- slight; 3- moderate to heavy
20. Pellicle on broth: 0- ND; 1- none; 2- ring; 3- thin, barely visible; 4- moderate to heavy
21. Maximum temperature at which growth is observed to occur: 0- ND; 2- 25°C; 3- 30°C; 4- 37°C; 5- 45°C; 6- 55°C

Physiology and Biochemistry (NC = No change or basic; A = acid; AG = acid and gas)

22. Glucose: 0- ND; 1- NC; 3- A; 4- AG
23. Fructose: 0- ND; 1- NC; 3- A; 4- AG
24. Galactose: 0- ND; 1- NC; 2- A; 4- AG
25. Lactose: 0- ND; 1- NC; 2- A; 4- AG
26. Sucrose: 0- ND; 1- NC; 3- A; 4- AG
27. Maltose: 0- ND; 1- NC; 3- A; 4- AG
28. Mannitol: 0- ND; 1- NC; 3- A; 4- AG
29. Sorbitol: 0- ND; 1- NC; 3- A; 4- AG
30. Arabinose: 0- ND; 1- NC; 3- A; 4- AG
31. Xylose: 0- ND; 1- NC; 3- A; 4- AG
32. Litmus milk: 0- ND; 1- not reduced; 2- reduced
33. Litmus milk: 0- ND; 1- no acid curd; 2- acid curd (coagulated)
34. Litmus milk rennet curd (no acid): 0- ND; 1- no rennet (score here if acid formed); 2- rennet curd formed
35. Blood agar: 0- ND; 1- no hemolysis; 2- beta hemolysis; 3- alpha hemolysis
36. Starch hydrolysis: 0- ND; 1- negative; 2- positive
37. Gelatin hydrolysis: 0- ND; 1- negative; 2- positive
38. Casein hydrolysis: 0- ND; 1- negative; 2- positive
39. Lipid hydrolysis: 0- ND; 1- negative; 2- positive
40. Nitrite from nitrate: 0- ND; 1- negative; 2- positive
41. Gas from nitrate: 0- ND; 1- no gas; 2- positive (Note: if positive, score No. 40 as 2)
42. Ammonia from peptone: 0- ND; 1- negative; 2- positive
43. Catalase: 0- ND; 1- negative; 2- positive
44. Oxidase: 0- ND; 1- negative; 2- positive
45. H_2S: 0- ND; 1- negative; 2- positive
46. Indole: 0- ND; 1- negative; 2- positive
47. Methyl red: 0- ND; 1- negative; 2- positive
48. Voges-Proskauer: 0- ND; 1- negative; 2- positive
49. Simmons citrate: 0- ND; 1- negative; 2- positive
50. Urease: 0- ND; 1- negative; 2- positive

Questions

1. All bacterial isolates belonging to a species would be expected to give the same result for a specific biochemical test. yes____ no____
2. Plasmids may convert a strain of bacteria either gaining or losing a characteristic depending on whether the plasmid is gained or lost. yes___ no___
3. Dichotomous keys do not always work, because all strains of a species may not have the same result for a particular test. yes____ no____
4. One can successfully use similarity coefficient taxonomic methods with a small number (fewer than 10) of characteristics. yes___ no___
5. The ideal for computer (Adansonian) taxonomy is to do as many tests on an organism or group of organisms as possible (i.e., 100 or more). yes___ no___
6. The process of arranging organisms into groups based on common characteristics is called ______________________.
7. A group of organisms with common characteristics is referred to as a ________________.
8. Giving a name to an organism is called ____________________.
9. When an unknown organism is compared to a group of known organisms, the process is referred to as ________________.

10–11. A number of methods for identifying organisms exist. The two described in this exercise are __________________ __________________ and __________________ __________________.

12–15. Ten characteristics are determined for an unknown organism and compared with data from known organisms. The data is given below. Calculate a %S for the unknown organism against each of the known organisms below. Using the cutoff from the exercise, determine which, if any, is identical to the unknown. (+ = positive, – = negative, 0 = missing)

Organism	Character									
	1	2	3	4	5	6	7	8	9	10
unknown	+	+	–	+	0	–	+	–	–	–
#1	–	+	–	+	–	+	–	–	+	+
#2	+	–	–	+	–	–	+	+	–	–
#3	–	–	+	–	–	–	+	+	+	–

12. $\%S_{sm}$ to #1 ________________
13. $\%S_{sm}$ to #2 ________________
14. $\%S_{sm}$ to #3 ________________

15. Is the unknown identical to any of these knowns using the cutoff of this exercise? Why?

Name ______________________________ Date ________ Grade ________

34A. Identification with a Miniaturized, Rapid Biochemical System: API 20E

Results and Observations

Unknown No. ____________

api 20 E

bioMérieux

1	2	4	1	2	4	1	2	4	1	2	4	1	2	4	1	2	4	1	2	4	1	2	4	1	2	4
ONPG	ADH	LDC	ODC	CIT	H_2S	URE	TDA	IND	VP	GEL	GLU	MAN	INO	SOR	RHA	SAC	MEL	AMY	ARA	OX	NO_2	N_2	MOB	McC	OF-O	OF-F

The codes below will identify six bacterial species, provided the results for the strains used are identical with the strains listed. If the lab strains have differences, the API Index will have to be consulted for proper identification.

Escherichia coli	5 144 552
Salmonella enteriditis	6 704 552
Klebsiella pneumoniae	5 215 773
Enterobacter aerogenes	5 305 773
Serratia marcescens	5 307 761
Proteus vulgaris	0 476 021

Questions

(Your instructor may select questions from Exercise 33 in addition if that exercise was not performed by the class.)

1. The ONPG test is a test for ____________ ____________________ .

2. A positive test for lysine decarboxylase results from the removal of a ______________ ______________ .

3. The citrate test is based on the ability to use citrate as a source of __________________ .

4. The VP test reagents are the same as used for the coliform _________ test.

5. The black precipitate in the H_2S cupule is ___________ ___________ .

6. A positive test on one of the carbohydrates depends on the production of ___________ .

7. The addition of mineral oil to certain tubes (e.g., H_2S) is intended to establish an ___________________ condition.

8–9. The *Enterobacteriaceae* are generally catalase _________ and oxidase ______________ .

10–11. In this exercise, zinc dust reduces ____________ to ___________ .

12. If an organism does not reduce nitrate, the addition of zinc dust will produce a ______________ test for nitrite.

Name ______________________________ Date ________ Grade ________

34B. Identification with a Miniaturized, Rapid Biochemical System: ENTEROTUBE II

Results and Observations

Unknown No. ____________

Record the results as + or – ***in*** the box of the test and ***circle*** the number below the box if the result was positive. Sum the circled numbers within the bracket and enter the number in the hexagon below the arrow. The five numbers reading from left to right make up the ID Value. Go to the Code Book and look up the number. Enter the name of the organism (or organisms) in the space provided and other data as indicated. **If you were supplied with a form, use that instead of the drawing below.** Attach the form to this page and submit both to your instructor.

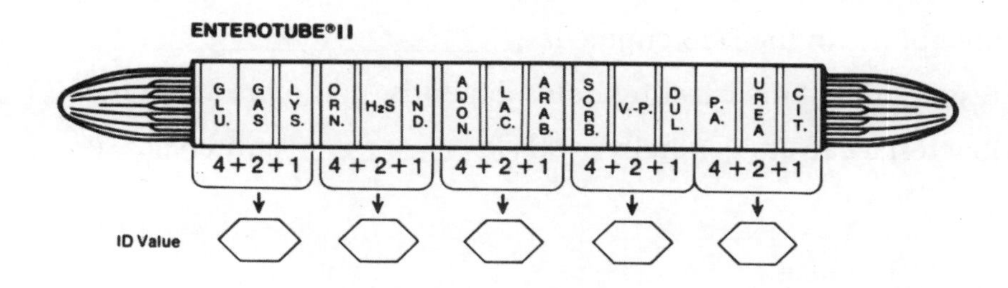

Organism name ______________________________

	Probability	Confirmatory Tests
Other organisms____________	____________	____________
____________	____________	____________
____________	____________	____________

Questions

(Your instructor may select questions from Exercise 33 in addition, if that exercise was not performed by the class.)

1. If your organism had an ID Value of 00001, what test should you perform? ____________
2. Name one oxidase-negative, glucose-nonfermenting organism: ________________ ________________

3–4. Name two tests determined in the API 20E kit not utilized in the ENTEROTUBE II kit.
3. ____________________
4. ____________________

5. A positive test for lysine decarboxylase results from the removal of a ____________ ________.
6. The citrate test is based on the ability to use citrate as a source of ________________.
7. The VP test reagents are the same as used for the coliform __________ test.
8. The black precipitate in the H2S cupule is __________ __________.
9. A positive test on one of the carbohydrates depends on the production of __________.

10–11. The Enterobacteriaceae are generally catalase-__________ and oxidase-__________.

Name ______________________________ Date ________ Grade ________

35. Water Quality Analysis: MPN Method

Results and Observations

Water Sample Source ______________________

Test	Sample volume (ml)							MPN
	10	10	10	10	10	1.0	0.1	
Presumptive LST Broth	/	/	/	/	/	/	/	XXXX
Confirmed BGLB Broth	/	/	/	/	/	/	/	
Fecal Coliforms EC Broth								
Completed LST Broth			+ = Gas – = No Gas / = 24 hr observations at upper left and 48 hr observations at lower right					
Gram Stain								

Conclusions:

Questions

1. MPN means ____________ ______________ ____________ of bacteria per milliliter.
2. A positive presumptive test consists of ________ production only.
3. The medium used to detect fecal coliforms is ______ ________________.
4. In disease, potential coliforms are normally considered to be ________________.
5. Drinking water is examined for the presence of coliform organisms mainly because they are normal ______________ inhabitants.
6. The step that proves an organism is a coliform is the ________________ test.
7. _______ broth is used in the coliform presumptive test.
8. The ______________ filter method also can be used to analyze drinking water.
9. The incubation temperature used for the MPN method is ______°C.
10. The Gram stain is a necessary part of the ______________ test.

11–14. List four waterborne diseases.

11. ______________________ 13. ______________________

12. ______________________ 14. ______________________

15. Why use coliforms instead of directly isolating pathogens from water?

____

16. The water analysis test is designed to detect the presence of ______________ bacteria of the ______________ group.

17–21. What is the definition of a coliform? (Analyze the coliform tests.)

22–24. List the 3 steps, in order, of the MPN test.

22. ______________________

23. ______________________

24. ______________________

25. ___________ agar is used to isolate typical coliforms for further study.

26–30. List 5 chemical agents used in the media for the MPN test that inhibit the growth of Gram-positive organisms while encouraging the growth of coliforms.

26. ______________________ 29. ______________________

27. ______________________ 30. ______________________

28. ______________________

Name ____ Date ________ Grade _________

36. Standard Plate Count of Food

Results and Observations

1. Record the number of bacteria observed in the table below.

Food	Dilution Counted	No. of Colonies on Plate	Total No. per gm
Hamburger			
Black pepper			
Chili powder			

2. Comment on colony morphology between foods.

3. Describe the Gram reaction and morphology of organisms from the three foods.

Questions

1. The term "colony forming units" means the same as the __________ of ____________.
2. Fresh foods usually have microbes on them. True ___ False ___
3. A spoiled food will often have ________ bacteria/g.
4. The total plate count is directly related to health hazard. True ___ False ___
5. If food-poisoning bacteria are present in a low number, such as 10^4/g, the food may still be _________________________.
6. A most important interpretation of numbers of bacteria over 10^5/g in food is that the food ____.
7. The medium used here for the total count is ___________ ___________ __________ agar.
8. An example of a food having microbial standards for quality and health is ____________.
9. The standard plate count (would/would not) be used on fermented foods.
10. Eleven grams of hamburger ground in 99 ml of dilution water is a _______ dilution.
11. The use of lots of chili pepper or curry in a food will make it safe to eat. True___ False___

Name ______________________________ Date ________ Grade ________

37. Standard Plate Count of Milk

Results and Observations

Milk Source ____________________ Raw ___ Pasteurized ___

Dilution	Reciprocal of Dilution	Colonies per plate	Average Colonies per plate	SPC/ml
10^0				
10^{-1}				
10^{-2}				
10^{-3}				

Questions

1. A colony count determines the numbers of ____________ bacteria in the milk sample.
2. A countable plate must have between _____ and _____ colonies.
3. A good-quality raw milk often has (a much greater/the same) number of bacteria as pasteurized milk.
4. The medium required for milk plate counts is _________ _________ _________.
5. Colonies are counted on milk plates using a ____________ ____________ ____________.
6. Milk is made safe to drink by a process called ______________________.

7–11. List 5 sources of potential contamination of milk.

7. ______________________ 10. ______________________

8. ______________________ 11. ______________________

9. ______________________

12. The ________ or ______________ ________ ________ procedure is used by the American Public Health Association to determine viable bacterial populations in milk.

13–14. Milk dilution plates are incubated at ____°C for _____ hours.

15. As a minimum, pasteurization methods are designed to kill the organism
______________ ______________.

16–17. According to public health law, acceptable milk for market must have less than ___________ bacteria/ml in pasteurized milk and less than ___________ bacteria/ml in raw milk.

18–20. Name 3 methods of pasteurization.

18. ______________________ 19. ______________________ 20. ______________________

21–23. List 3 diseases of humans introduced into milk by uncleanliness of the animals, equipment, etc.

21. ______________________

22. ______________________

23. ______________________

24–26. List 3 diseases transmitted to humans by drinking milk from infected cows.

24. ______________________

25. ______________________

26. ______________________

Name ______________________________ Date ________ Grade ________

38. Preparation and Analysis of Yogurt

Results and Observations

A. Description of Yogurt

Odor ______________________________
Texture ______________________________
Taste ______________________________

B. Microscopic Count

Cell Type	Average No./Field	Dilution Factor	Microscope Factor	Count/gram
Cocci				
Rods				
Total				

Description of the morphological forms of bacteria observed:

C. Viable Count

Colony Type	Average No. of Colonies	Dilution Factor	Count/gram	Gram Reaction	Catalase Reaction
Total					

Colony descriptions:

Questions

1. The principal acid produced in yogurt is ________ ________.

2. In heating the milk for making yogurt, care must be taken to avoid __________.

3. The optimum yogurt incubation temperature is ____°C.

4. Once made, a yogurt inoculum is good for a (short/long) period of time.

5. Adjusting the wattage of a _______ _______ permits the use of an insulated picnic hamper as an incubator.

6. A candle jar will have a higher concentration of ________ (abbreviation) than normal air.

7. *Streptococcus thermophilus* colonies are (describe) ____________, ______________, and _______________.

8–9. The coccus member of the yogurt-making pair is named ____________________ __________________ and the rod-shaped member is ______________________ __________________.

10–11. Both bacteria involved in yogurt making are Gram-____________ and catalase-____________.

12. The catalase test is performed using ______________________ as an enzyme substrate.

13–14. The coccus member of the yogurt pair is a ________-lactic fermenter and the rod is a ________-lactic fermenter (see *Bergey's Manual*).

15. The medium used to grow the yogurt bacteria is specially formulated for that purpose and is called _________ agar.

16. Colonies of the rod-shaped bacteria are ___________, _________, and ______________.

17–18. The purpose of the candle jar is to increase _________ _________ and decrease _________.

19. Spoilage of yogurt is most likely to occur by ___________ (group of organisms). (See Exercise 3 for clues.)

Name ________________________________ Date ________ Grade ________

39. Replica Plating for a Nutritional Mutant

Results and Observations

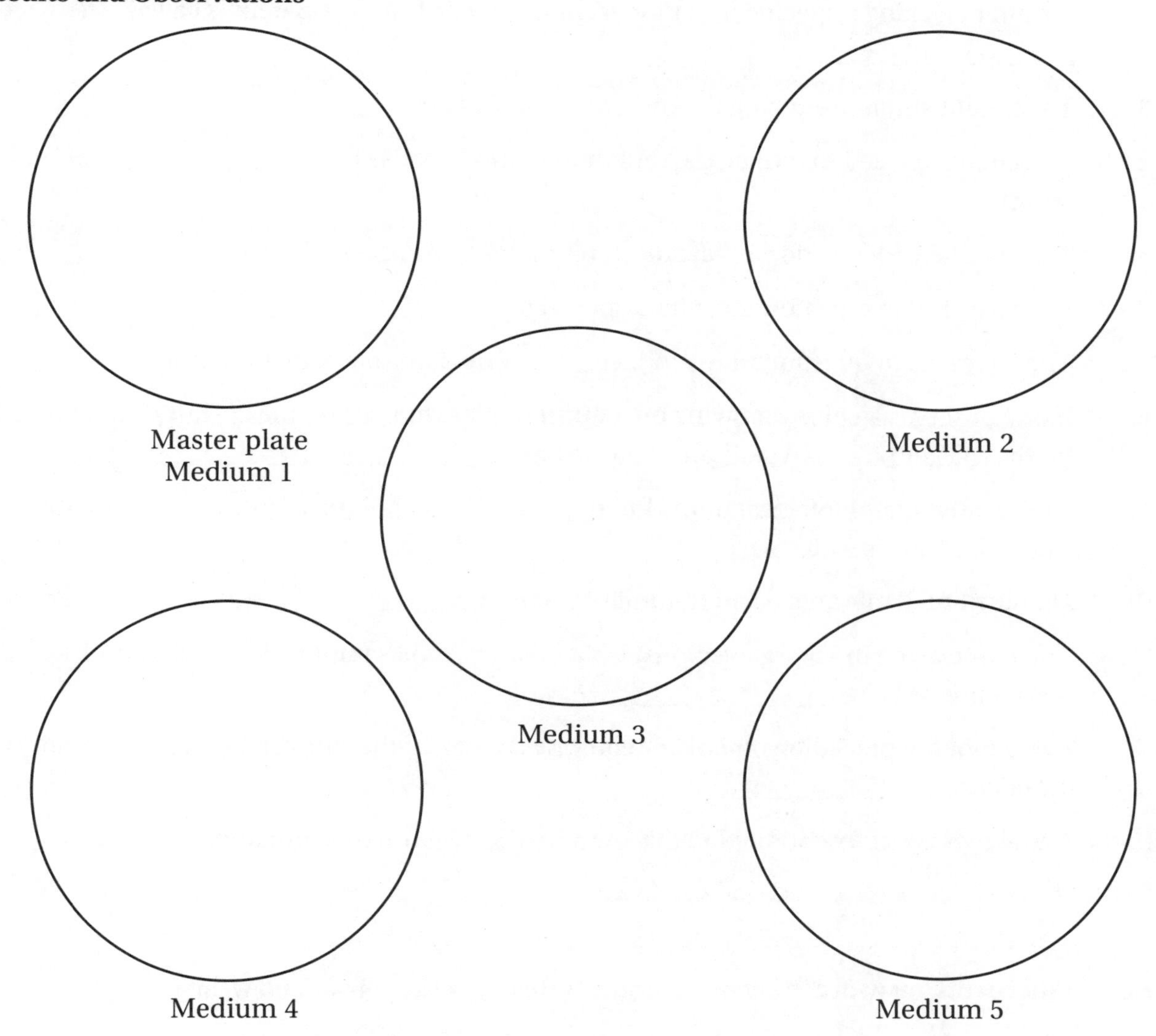

Verification:
On the form above, mark the colony selected for verification with the letter **S**.

Growth on medium 1 _______ (use + or –)

Growth on medium 2___ 3___ 4___ 5___ (use + or –)

Conclusions:

Questions

1. In this exercise, the most likely DNA mutation to occur is the replacement of ____ with ____.
2. A mutant lacking a specific mutational characteristic from the parent is called a(n) ________________.
3. The parent strain producing the mutation is called a(n) ________________.
4. The chemical used to induce the mutation in this exercise is ________________ (abbreviation).
5. The chemical in question is hazardous to humans because it can cause ____________.
6. The chemical in question must be disposed of as a ______________ ______________.
7. The chemical in question may produce ____ % yield of a particular mutant.
8. If one observes a colony growing on a nutrient plate but not on the corresponding spot on the master plate, it would probably be an ____________________.
9. As in many microbiological procedures, _________ technique is critically important in this procedure as well.
10. Do nutritional mutants occur naturally? yes ___ no ___
11. The name given to an organism just isolated from a soil sample relative to its nutritional requirements is ________________.
12. A strain of the preceding organism requiring a nutrient that the original strain did not is called a(n) ______________.

13–14. Name two agents mentioned in this exercise that will produce mutants.

13. ______________________________
14. ______________________________

15. When working with the chemical agent of this exercise, you should wear ___________.
16. The technique used to test a large number of strains for nutritional requirements is ______________ ________________.

Name ______________________________ Date ________ Grade ________

40. Bacterial Transformation

Results and Observations

1. Record your results using the format suggested in the kit.

Conclusions and discussion:

Questions

1–2. Two terms are used in genetic engineering to mean an increase in the number or amount of a plasmid or gene. These two terms are ________________ and ________________.

3–4. The plasmid into which a gene has been inserted in the laboratory is called a ____________. The process of inserting the gene is called ____________.

5. A host is called competent if it can ______________ a plasmid.

6. A small number of host cells in a culture can be made competent by treating the culture with the chemical __________ (chemical formula).

7. The uptake of plasmid DNA by a competent cell in the laboratory is called ________________________.

8. In the case of the amplification of an ampicillin resistant plasmid, what is an essential ingredient of the growth medium? ________________

9–10. Name two other factors that may be part of a plasmid

9. ________________________

10. ________________________

11. A plasmid with a newly inserted gene is referred to as a __________________ plasmid.

Name ______________________________ Date ________ Grade ________

41. Normal Flora of the Human Throat

Results and Observations

Colony Description	Size mm	Chromogenesis	Type of Hemolysis	Gram Reaction	Morphology

Questions

1–2. "Strep throat" may be caused by ________________ ________________ and is ________-hemolytic.

3–4. The ________________ ________________ are the most prominent organisms of normal throat flora throughout life and they are ________-hemolytic.

5. If a throat swab cannot be cultured immediately, it should be placed in a tube of ________________ ______________ to prevent drying.

6. The numerical predominance of a specific organism in a throat culture usually indicates that it (is/is not) responsible for a pathogenic condition in the throat.

7–10. Name four indicators useful in the identification of an organism from a throat culture.

7. ______________________ 9. ____________________

8. ______________________ 10. ____________________

11. Hemolysis is produced as a result of changes in ______________________.

12. Pathogenic streptococci are generally ________-hemolytic.

13–16. List four organisms or genera that are part of the normal flora of the human throat.

13. ______________________ 15. ______________________

14. ______________________ 16. ______________________

17–19. __________ ____________ is the differential medium used to culture the throat specimen with incubation at ____°C for ____ hours.

Name ______________________________ Date ________ Grade ________

42. Dental Caries Susceptibility

Results and Observations

Sample	Incubation Interval			Degree of Caries Susceptibility
	24 hr	48 hr	72 hr	
Control				
Saliva				

+ = positive
– = negative

Questions

1. The pH indicator in the Snyder test agar is _____________ _____________.

2. A positive reaction with the pH indicator in the Snyder test agar is when it turns _____________.

3. Interpretation of the Snyder test results depends only on the amount of _________ produced.

4. Tooth decay generally begins at pH _____ and below.

5. _________________________ are fermented to organic acids to produce tooth decay.

6. The acid believed to be responsible for dental caries is __________ acid.

7–9. _____________________ _________is the primary organism involved in dental caries. It is a Gram-___________, non-sporeforming ___________ (morphology) organism.

10–11. List the two main ingredients in Snyder test agar.

10. _________________________________

11. _________________________________

12–15. List four measures that in your estimation will prevent dental caries.

12. _________________________________

13. _________________________________

14. _________________________________

15. _________________________________

16–17. Do the results of this exercise agree with your case history of dental caries? Explain.

____

____

____

Name ______________________________ Date ________ Grade ________

43. Handwashing and Skin Bacteria

Results and Observations

Bowl		Count/volume			Count/ml in Bowl	Count/hand
		1 ml	0.5 ml	0.1 ml		
No. 1	PCA					
	MSA					
No. 2	PCA					
	MSA					

Morphology and Gram stain

1.
2.
3.
4.
5.
6.
7.
8.
9.
10.

What was the percent reduction in total count between the two washings? ____________%

$$\frac{\text{(PCA count/ml bowl 1)} - \text{(PCA count/ml bowl 2)}}{\text{(PCA count/ml bowl 1)}} \times 100 = \%$$

Was there a change in morphological types and/or Gram reaction between the bowls? If so, what was the change?

Questions

1. Bacteria that are easily removed from the skin are called the ____________ bacteria.
2. Bacteria normally living in or on the skin are called _____________ bacteria.
3. Handling a chicken with salmonellae on the skin could contribute to the __________________ bacteria on your hands.
4. The man who first observed the relationship between handwashing and disease was ________________ ______________________.
5. The disease he helped reduce by requiring medical students to wash their hands was ________________ ______________.
6. The disease in Question 5 was caused by __________________ __________________.
7. It (is/is not) common to find coliforms on the hands.
8. MSA medium is selective for ________________________.
9. In addition to the bacteria, ___________ may be found in the resident microbe group.
10. Mold spores may be part of the ______________ flora.
11. Typhoid fever may be spread by hands in contact with ___________.
12. Virus diseases (would/would not) be expected to be transmitted by hands.
13. The resident bacteria are usually (harmless/virulent).
14. The Gram-_____________ bacteria are usually more common among the resident bacteria than the transient group.
15. Generally speaking, the _______________ microbes are most important in medical procedures.

Name ______________________________ Date ________ Grade __________

44. Culture and Examination of Urine and Blood

Results and Observations

I. Urine Specimen

1. Gram stain results:

   a. How many different types of organisms were seen? ________

   b. What was the Gram stain reaction, cell shape, and cell arrangement?

   ____

   ____

   c. If more than one organism was present, which one was the most numerous?

   ____

2. Urine culture results:

   a. How many different colony types were seen on the blood agar plate? __________

   Describe the colony morphology ______________________________

   ____

   b. What is the Gram stain reaction, cell shape, and cell arrangement of the growth on the blood agar plate? ______________________________

   ____

   c. How many different colony types were seen on the EMB agar plate?

   ____

   d. What is the Gram stain reaction, cell shape, and cell arrangement of the growth on the EMB agar plate? ______________________________

   ____

   e. Do the results of the urine culture plates agree with the results of the direct Gram stain? ______________________________

   ____

II. Blood Specimen

1. Gram stain results:

   a. How many different types of organisms were seen? ________

   b. What was the Gram stain reaction, cell shape, and cell arrangement?

   ____

   ____

   c. If more than one organism was present, which one was the most numerous?

   ____

2. Blood culture results:

a. How many different colony types were seen on the blood agar plate? __________

Describe the colony morphology ____

b. What is the Gram stain reaction, cell shape, and cell arrangement of the growth on the blood agar plate? ____

c. Do the results of the blood culture plates agree with the results of the direct Gram stain? ____

Questions

1. Name the organism that is the most common infecting agent in urinary tract infections. ______________ ______________

2–4. Name three organisms or genera that are part of the normal flora of the urethra.

2. ____________________ 3. ____________________ 4. ____________________

5. Normal urine is _____ of bacteria.

6. Cystitis is an inflammation of the _______________.

7. When urinary infectious agents travel up the ureters to the kidneys, the inflammation of the kidneys is called ______________________.

8. EMB is a mildly selective medium for Gram-__________ organisms.

9–10. The two dyes in EMB responsible for the selective effect are __________ and ________________ __________.

11. If a bacterium forms a pigmented colony on EMB agar, this suggests that it is capable of fermenting the carbohydrate ___________.

12. Normally, blood is _______ of bacteria.

13. The viridans group of streptococci are _______-hemolytic.

14. In blood culturing, the proper ratio of blood to broth is ___to___.

15. Blood agar usually allows growth of _____ organisms.

16. Name the organism that is the most common cause of subacute bacterial endocarditis. _________ ___________________.

17. In the hospital, culture plates of patient specimens are routinely processed by medical ________________________.

18. The processing of culture plates in the hospital, even if automated, requires human ____________________.

19–20. Name two professionals who work in the clinical laboratory and process patient cultures.

19. ________________________ 20. ________________________

Name ____ Date ________ Grade __________

45. Antimicrobial Susceptibility Testing

Results and Observations

Use R (resistant), I (intermediate), or S (sensitive) from Table 45-1 in the small box under the species name.

Antibiotic	Disk conc.	Inhibition Zone Diameter (mm)							
		S. aureus		*E. coli*		*P. aeruginosa*		*S. cerevisiae*	

Other Observations:

Questions

1. A single antibiotic is usually toxic to both prokaryotes and eukaryotes. True __ False __

2. An antibiotic is usually effective in __________ doses.

3. Standard strains of bacteria are used as ____________ in the Kirby-Bauer method.

4. If the zone size with the standard strain varies, it often means the test was performed __________________ than it should have been.

5. Antiseptics and antibiotics are (different/the same).

6. The name of the standardized disk test for antibiotic sensitivity is the _____________ ______________ disk test.

7–10. Name four factors that may affect the diameter of the inhibitory zone in any particular test.

   7. ______________________________
   8. ______________________________
   9. ______________________________
   10. ______________________________

11. The name of the medium used in this test is ________________ ________________ agar.

12–14. Name the three microbial genera responsible for producing most of the antibiotics.

   12. ______________________________
   13. ______________________________
   14. ______________________________

15. In the space below explain why *Saccharomyces cerevisiae* was not affected by these antibiotics.

Name ______________________________ Date ________ Grade ________

46. Action of Disinfectants and Antiseptics

Results and Observations

Disinfectant/ Antiseptic	Conc'n	Inhibition Zone Diameter (mm)			
		S. aureus	*E. coli*	*B. subtilis*	*S. cerevisiae*

Other observations (Note the chemical and organism):

Questions

1. Sanitizers are used on only on __________ utensils.
2. Zones of inhibition are measured and recorded in ____________________ .
3. A disinfectant is expected to let endospores ______________ .
4. An antiseptic is usually used on ________________ objects.
5. An agent used as a gargle would commonly be called a(n) ____________________ .
6. The "zone of inhibition" refers to the clear area around a chemical agent disk in which __________________________________ .
7. The ______________ __________________ _______ is a regulatory test to measure the effect of a disinfectant on pathogens.
8. ____________________ are agents used to kill bacteria on well-cleaned food handling equipment.
9. The suffix -________ means "to kill."
10. The suffix -________ means "to arrest," "to stop," or "to inhibit".
11. Agents used to kill or inhibit growth of vegetative cells are generally called ____________________________ .

12–13. List two major differences between an antibiotic and a disinfectant.

12. __________________________________
13. __________________________________

Name ____ Date _________ Grade __________

47. Agglutination Reactions: Blood Grouping and the Rh Factor

Results and Observations

Blood type _____________

Rh factor ______________

Questions

1. Do you think that outdated, old reagents might result in a false positive blood typing? yes __ no __
2. The factor on the red blood cell that gives an Rh positive test is an _________________ .
3. An individual with the blood type A has naturally occurring antibodies against blood type ___.
4. Human red blood cells are antigenic only in other individuals of (a different/the same) blood type.
5. About _____% of the human population is Rh negative.
6. The typing serum for blood type A can be obtained from the serum of a person with blood type _____.
7. Agglutinogen is synonymous in meaning to the term _______________ .

8–9. The two typing sera used to determine all four blood types are called __________-_____ and ____________-_____ .

10. The amount or level of antibodies in an antiserum is often referred to as ____________ .

Name ______________________________ Date __________ Grade __________

48. Enzyme-Linked Immunosorbent Assay (ELISA)

Results and Observations

Record the results as + or – for purple color. Note the highest dilution showing purple.

	Well No.											
	1	2	3	4	5	6	7	8	9	10	11	12
Antiserum												
Control												

The antiserum titer is ______________

Questions

1. ELISA stands for ____.
2. The acid color of bromcresol purple is ____________ .
3. The enzyme substrate used in this exercise is ___________ .
4. Enzyme immunoassays are considered to be (more/less) sensitive than agglutination reactions.
5. An alternative to the use of an enzyme in these assays is ________________________ .
6. The amount of antibody present in a volume of serum is referred to as the ___________ .
7. A most important step between the addition of reagents to the wells is a thorough ______________ of the wells.

8–9. The products of urea hydrolysis are ________ and ______________________ .

10–13. Name four enzymes conjugated to antibodies used for enzyme immunoassay.

10. ________________________ 12. ________________________

11. ________________________ 13. ________________________

14–17. List the order of reagent application in the test.

14. ________________________ 16. ________________________

15. ________________________ 17. ________________________

18–19. The antirabbit IgG conjugated to the urease was made in ___________ (animal), while the albumin antibody was made in _______________ (animal).